# ARCHITECTURE REPERFORMED:
# THE POLITICS OF RECONSTRUCTION

# Architecture RePerformed:
# The Politics of Reconstruction

Edited by
Tino Mager
*Berlin Institute of Technology, Germany*

LONDON AND NEW YORK

First published 2015 by Ashgate Publishing

2 Park Square, Milton Park, Abingdon, Oxfordshire OX14 4RN
711 Third Avenue, New York, NY 10017

*Routledge is an imprint of the Taylor & Francis Group, an informa business*

First issued in paperback 2017

**British Library Cataloguing in Publication Data**
A catalogue record for this book is available from the British Library.

**Library of Congress Cataloging-in-Publication Data**
Architecture RePerformed : The Politics of Reconstruction / [edited] by Tino Mager.
        pages cm
    Includes index.
    ISBN 978-1-4724-5933-6 (hardback : alk. paper)
    1. Architecture and society. 2. Architecture – Political aspects. 3. Buildings – Repair and reconstruction. 4. Reproduction (Psychology) I. Mager, Tino, editor.

    NA2543.S6A6345 2015
    724'.6 – dc23

                                                                  2015014506


ISBN 978-1-4724-5933-6 (hbk)
ISBN 978-1-138-57329-1 (pbk)

# Contents

# List of Figures

# Notes on Contributors

## DR ARNOLD BARTETZKY

Research Coordinator for Art History at the Centre for the History and Culture of East Central Europe at Leipzig University (GWZO). Sessional teaching of History of Art at the Universities of Leipzig, Jena and Paderborn. Regular contributor to the cultural section of the German newspaper *Frankfurter Allgemeine Zeitung*. Principal fields of research: architecture, city planning, monument preservation and political iconography from the nineteenth century to the present, architecture of the Renaissance and the Mannerism.

## DR JULIEN BASTOEN

Associate teacher in Architectural History at the Ecole Nationale Supérieure d'Architecture de Paris La Villette. PhD in architecture at University Paris Est. Writes as architecture critic for *Criticat* and *Archiscopie*. Fellow at Keepers Preservation Education Fund. Various articles on architectural cloning, museum and department store architectural history.

## DR ROBERT BORN

Research Fellow at the Centre for the History and Culture of East Central Europe at Leipzig University. Sessional teaching of history of art at the Universities of Leipzig, Basel and Berlin. Specialized in historiography of art and nationalism in South-Eastern Europe, symbolic constructions of the past in South-Eastern Europe.

## DR RENATO CYMBALISTA

Architect and Urbanist, PhD in Architecture and Urbanism (University of São Paulo, Brazil, 2006), Post-doc in History at the University of Campinas. Professor of Urban History at the School of Architecture and Urbanism of the University of São Paulo. Coordinator of Urbanism of Instituto Pólis, a São Paulo-based NGO and research Institute. Visiting scholar and visiting professor in several universities and institutions including ISCTE Lisbon, Universidade Nova de Lisboa, Brandenburgische Universität Cottbus and the John Carter Brown Library, Providence.

## DR JOSEP-MARIA GARCIA-FUENTES

Lecturer in Architecture at Newcastle University (School of Architecture, Planning and Landscape). He is also an architect, Fellow of the London School of Economics – Catalan Observatory – and Adjunct Professor at the Department of Architectural Composition at the Universitat Politècnica de Catalunya-BarcelonaTECH. Formerly vice dean and international coordinator at the Escola Tècnica Superior d'Arquitectura del Vallès-Barcelona (2011–14) and Assistant Professor (2010–14) in the same university. Winner of the First National Prize of Spain for university graduates 2006. Fellow by the Caja de Arquitectos (2004), the Universitat Politècnica de Catalunya-BarcelonaTECH (2007), the Ministry of Science and Innovation of Spain (2007–10) and the Samuel H. Kress Foundation for the Society of Architectural Historians (2011).

## DR ALEXANDRA KLEI

Lecturer in Art History at Ruhr University Bochum. Current research topics: architect Hermann Zvi Guttmann and the conditioning for Jewish building developments in Germany after 1945, 'White City' Tel Aviv (Habilitation treatise), architecture and town planning in Israel, memorial places, architecture after 1945, photography.

## JOÃO CARLOS SANTOS KUHN, MA

Architect and Urbanist (University of Brasília), Masters in Architecture and Urbanism, School of Architecture and Urbanism (University of São Paulo), researcher in the research group 'Sites of memory and Conscience' (University of São Paulo/ National Counsel of Technological and Scientific Development, Brazil).

## DR TINO MAGER

Art and Architectural Historian. Studies in Media Technology at Leipzig University of Applied Sciences (HTWK), graduate engineer. Studies in Art History and Communication Sciences at Berlin Institute of Technology, University of

Barcelona (Spain) and Sophia University Tokyo (Japan), MA. PhD in art history at Berlin Institute of Technology. Dissertation on the notion of authenticity regarding historical architecture. Visiting Researcher at University of California, Los Angeles. Elsa-Neumann Fellow. Lectureships at Berlin Institute of Technology and Istanbul Technical University (Turkey). Various publications on transnational artistic education, twentieth century architects and traditional Japanese architecture. Currently a postdoctoral research fellow at research group WDWM (Which Monuments, Which Modernity?) at TU Dortmund University / Bauhaus University Weimar.

## DR ALICE Y. TSENG

Associate Professor, History of Art and Architecture, at Boston University. Fellowships from numerous institutions and foundations, including the Fulbright Foundation, Center for Advanced Study in the Visual Arts (National Gallery of Art), J. Paul Getty Foundation, Metropolitan Center for Far Eastern Art Studies, American Council of Learned Societies. Recipient of the 2006 Founder's Award from the Society of Architectural Historians. Specific topics of research interest are the history of institutional buildings, collections, exhibitions, and transnational and transcultural connections between Japan and Euro-America.

## DR JING ZHUGE

Associate Professor at the School of Architecture, Southeast University, China. She graduated in architectural history and theory from the Southeast University in 1997 and also completed her PhD there in 2004. Subsequently, she worked as a professor in the School of Architecture at the university. She teaches the History of Chinese Architecture, Methodologies of Architectural History Research, and Surveying and Measuring of Architecture to graduate students. Her main fields of research are the social significance of architecture in history, the historiography of Chinese architecture and the reconstruction of traditional Chinese concepts of architecture.

# Introduction: Selected Pasts, Designed Memories

*Tino Mager*

Of all the architectural movements that emerged in the twentieth century, reconstruction has proved the most successful. What sounds odd at first turns out to be plausible on taking a closer look at the facts. Despite the phenomenon of reconstruction being poorly investigated, no other recent architectural trend can claim comparable persistence along with global validity. Reconstruction, understood as the facsimile replication of lost buildings, emerged as a novel way of dealing with loss and evolved into a popular manner of adopting the past. Its significance can hardly be overestimated as we find reconstructions among highly prestigious current building projects: Moscow's Cathedral of Christ the Saviour, the Berlin City Palace and the Gyeongbok Palace in Seoul are but a few examples that represent the meticulous reconstructions of bygone architectural works that have been resurrected after being either absent for decades or replaced by other buildings.[1] These are landmarks at representative places in capital cities and they possess a tremendous power for the building of identity. However, they do not use the current architectural language and, hence, do not seem to express contemporary political, social and cultural ideals. Instead, these reconstructions restage history from the perspective of those in power today.

On examining the built environment of various cosmopolitan cities, we have to acknowledge that, especially during the last two decades, reconstruction has become an established way of building and simultaneously engaging with the past (for example Beijing, Dresden, Moscow, São Paulo and Warsaw). In many cases, even twentieth-century buildings are being replaced by reconstructions of their historic predecessors. Does this imply a failure of contemporary architecture? The politics of reconstruction go far beyond aesthetic considerations. Taking architecture as a major source of history and regional identity, the impact of large-scale reconstruction is deeply intertwined with political and social factors. Furthermore, memories and associations correlated with the lost buildings of a bygone era are heavily influenced by their reappearance, something that often contradicts historical events. Moreover, architectural reconstruction disintegrates the historical relations of the original building. Thus, connotations related to historic buildings can be replaced or manipulated by a huge influence on the society's collective memory.

Even among UNESCO World Heritage sites, we find full-scale reconstructions of completely destroyed buildings. The Historic Centre of Warsaw, Mostar's Old Bridge and the Tombs of the Buganda Kings at Kasubi are nothing but replicas of destroyed works of architecture. In talking about heritage, we touch upon history, a legacy and a testimony of the past. Historical architecture helps to understand and learn about times gone by. It assures us that the past was real and allows us to get a hold of the past. Yesteryear's architecture – be it from medieval times, the nineteenth century or the year 2000 – determines the appearance of our cities and villages; we live and work in it, and are familiar with its modes of representation and approaches to spatial organization. It is a vivid part of the past, deeply intertwined with the present. Therefore, architecture's power to design our image of the past can hardly be underestimated. It serves as the settings in our imaginations for historic events and demonstrates the ideals and potentials of our ancestors. Unfortunately, buildings, like all cultural achievements, are doomed to decay. Whether caused by neglect, war or catastrophes, architecture 'carries within itself the traces of its future destruction, the future perfect of its ruin' (Derrida 26). The associated losses are difficult to come by and attempts to undo them are comprehensible. But what does architectural heritage actually testify to, even world heritage, when it is merely a replica? By replication, architecture turns from a source of history into a result of our knowledge of history and loses its ability to provide a reliable account of the past. Moreover, in the case of buildings that vanished generations or even centuries ago, why do we compensate for a loss that we never experienced?[2] Where does this anxiety for compensation come from, and how does it re-enter collective memory?

The aim of this book is to take a closer look into the diverse methods of reconstruction, its historical development, reasons and effects. It provides a basis for the examination of a current yet scientifically underrepresented architectural trend.[3] By focusing mainly but not exclusively on the meticulous replication of lost buildings, it also touches on the field of hypothetical reconstruction as well as the reconstruction of urban identities. Critically investigating the subject and posing inconvenient questions should not be understood as a polemic for or against reconstruction, but as an invitation to further profound examination of the phenomenon. Nine chapters by internationally recognized specialists analyse the scope and meaning of reconstruction. By providing case studies from various countries, the book underlines the subject's global validity as well as the comparison of specific motivations. Furthermore, by covering a period of roughly a century, it allows for retracing the evolution of the backgrounds and purposes. Initially, a historical review will help understand why the current phenomenon of reconstruction differs from former ways of handling historical architecture.

## ASSIMILATION OF THE PAST

Since ancient times, buildings have survived only if they fulfilled a purpose, were needed and could be used. Many antique and later constructions became subjects of reuse if their original objective ceased.[4] The Parthenon in Athens, a temple dedicated to Athena for almost a thousand years and the ideal conception of antique architecture, became a Christian church in the sixth century, an Islamic mosque in the fifteenth century, a gunpowder magazine by the end of the seventeenth century and subsequently – as a result of its inadequacy for the latter use – a ruin. Its rediscovery took place only in the 1830s and attempts to bring back its original glory are still in progress. Many highlights of world architecture share a comparable fate. They were reused, modified, ruined, forsaken or forgotten until they entered the focus of a novel and intense interest for history in the nineteenth century.[5]

When buildings in use were damaged or destroyed, they were usually either repaired or rebuilt in the style of contemporary architecture to fulfil current modes of representation and up-to-date methods of construction.[6] The interest in regaining the historical dimension of buildings that experienced changes and damage over time emerged around 1800. This happened in alignment with a new awareness of and an interest in history.[7] One of the earliest examples for the novel interest in the architectural past – a practical examination that would soon be named 'restoration' – is Salisbury Cathedral in England. This thirteenth-century building was 'restored to its primitive simplicity and beauty' by James Wyatt between 1789 and 1792, whereby its additions over time, considered as 'defects' (Storer m), were removed. The objective was to obtain a stylistically purified building that represented pure Gothic architecture and was free from the disruptive elements of later times, such as Baroque altars, perpendicular chapels or the freestanding bell tower. John Milner, the later Vicar Apostolic of the Midland District, is among the first to express his regret about the restoration and loss of 'monuments of antiquity' (Milner 16). He criticizes the stylistic 'corrections' and condemns the creation of a merely contemporary idea about the past. Fearing that this treatment 'will extend itself to the remaining Cathedrals, and that there will not be a genuine unadulterated monument of sacred Antiquity left in this Island' (Milner 51), he anticipates the result of a movement that will soon spread to continental Europe.[8]

Undeniably, the nineteenth century witnessed affection towards historical architecture that was hitherto unknown. Notably, the re-establishment of Gothic architecture, supported by its usage for nationalistic ideas, led to an excessive building activity.[9] After the British, the French and, finally, German churches and castles became the subjects of restorations that focused on the purification and unification of their stylistic features. Countless historical buildings were modified to accord with nineteenth-century ideas of medieval architecture, which were essentially lacking profound knowledge and detailed research. The growing fascination with history in the nineteenth century rests largely on an incomplete and mystified image of the past, fuelled by a sensed alienation caused by industrialization and urbanization.[10] The recreation of the past, therefore, reflects the ideals and the longing of the time. This approach is revealingly expressed in the words of Eugène Emmanuel Viollet-le-Duc, France's leading Gothic Revival architect, when he defines 'restoration' in his *Dictionary of French Architecture* (1866) as 'to re-establish a building to a completed state, which may actually never have existed at any given time' (Viollet-le-Duc 14).[11]

It was again in England, where the protest against the restoration of historic buildings arose. The opponents considered the means as barbaric assaults against remains of the past, remains that they recognized as irrecoverable. Old buildings were defined by them as valuable and dignified heritage, as a legacy that belongs not only to the present but also to the generations to come. They demanded the preservation of the buildings to allow them to truly speak of the past whereas the restorations merely speak of the contemporary idea about the past. When John Ruskin, the major English art critic of the Victorian era, concludes: 'the greatest glory of a building is not in its stones, nor in its gold. Its glory is in its Age' (Ruskin 186), he underlines the acknowledgement of the significance of architecture's historic dimension. The opponents' late triumph is commonly seen in the case of Heidelberg Castle. The ruin of a Renaissance structure, destroyed in 1689 and 1693, came into the focus of a reconstruction campaign during the last third of the nineteenth century.[12] After the restoration of the burnt-out Friedrich wing, the reconstruction of the Ottheinrich wing became the subject of an extensive debate on conservation principles during which art historian Georg Dehio, one of the main adversaries of the reconstruction, coined the maxim 'conserve, do not restore!' (145).[13] After years of controversy, it was decided to leave the ruin a ruin, a highly representative decision for modern principles of monumental preservation. By the end of the nineteenth century, the first monument protection laws were already in effect. Outside of Europe, Japan was one of the first countries to adopt such a law (1897).[14] Sadly, the intellectual achievements concerning heritage conservation that were made during the century turned out to be unserviceable in consideration of the events to come.

## THE BIRTH OF ARCHITECTURAL RECONSTRUCTION

It only took a few seconds. After the incredible noise fell silent on this fateful Monday morning in 1902, the immense cloud of dust started to dissolve and slowly unveiled Venice's St Mark's Square, as it has never been seen before. The venerable Campanile that dominated the appearance of the square since the ninth century was gone. The millennium-old landmark had transformed into a mountain of rubble. All at once, there was neither the romanticization nor the completion of the past that accompanied the contemplation of ruins and stimulated their restoration; instead, there came a concrete and sore feeling: loss.[15] The difficulties caused by dealing with it led to an immediate decision: the prompt reconstruction of the tower.

For the first time, the precise rebuilding of a lost monument was initiated. Until then, meticulous reconstruction took place only on paper, in drawings or texts.[16]

*Wien VI.    Alte und neue Laimgrubenkirche.*

I.1    Vienna: Old and new Laimgrubenkirche, 1907
*Source:* Free Art License (http://en.wikipedia.org/wiki/Free_Art_License):
http://commons.wikimedia.org/wiki/File:LaimgrubenkircheAltNeu.jpg.

Now, however, it was decided to erect a facsimile of the Campanile, without plans, but at least based on photographs.[17] Ten years after the collapse, the tower has been (re)opened. The external appearance is much the same, but inside there is an elevator now and the tower's centuries-old stones have given way to a ferroconcrete structure with brick facing (Fenlon 145). Otto Wagner, the leading Austrian architect at the time, accused the reconstruction as a 'desire to falsify the history of architecture' (Pittarello).[18] Thus, from the outset, reconstruction is confronted with the allegation of being merely a copy, a forgery or fraud, since it lacks all historical depth and fools the spectator by pretending to be ancient and genuine. On the other hand, reconstruction is discovered as an effective instrument of coping with loss and enhancing urban situations. In Vienna, for instance, the Baroque church of St Josef ob der Laimgrube was copied because it was blocking traffic. Right behind the choir of the seventeenth-century building, an identical church was built, albeit smoother and neater, but the overall features were virtually the same (Figure I.1). Before the original building was torn down, both churches stood next to each other for a couple of months (Pfarre-St.-Josef-ob-der-Laimgrube). Today, nothing indicates that the apparently Baroque church is from 1907.

By then, unknown devastations caused by the World Wars brought new challenges with respect to the preservation of monuments as well as urban planning. The call for mere conservation became absurd in consideration of the plethora of monuments left in ruin by modern war technology. Completely erased cities brought up the question of how to tie in with their history and how to retain their identity. Ypres in Belgium and the Polish capital Warsaw are two successful examples of reconstructions of not only single buildings but also entire town centres. Both cities were almost entirely destroyed in the First and Second World Wars. At first sight, nothing indicates this immense break today. Both cities seem to look back on an uninterrupted historical continuity. Their resurrection allowed the residents to master the catastrophe and readopt the achievements of their past. However, cities like Rotterdam – completely rebuilt in contemporary forms after its destruction by German and American air raids in 1940 and 1943 – demonstrate that there are different ways of successfully overcoming disaster.

Whereas the reconstruction of buildings seems to be understandable in some situations, it appears difficult to consider them as identical with the lost buildings. This was already understood when the *Venice Charter*, the fundamental international framework for preservation and restoration, was adopted in 1964. For historical monuments, it demands 'to hand them on in the full richness of their authenticity' (ICOMOS 1971 LXIX). Furthermore, it excludes reconstruction as a valid means for preservation.[19] The *Venice Charter* is the cornerstone of the UNESCO *World Heritage Convention*; therefore, it defines the framework for the characteristics of World Heritage.[20] Nevertheless, reconstructions made their way into the World Heritage List time and again. Increasing numbers of charters and declarations concerning architectural preservation have loosened the strict attitude towards reconstruction.[21] At this point, we have to ask what architecture and architectural heritage really is. Buildings clearly emerged from pure necessity. But in all its diverse characteristics, architecture is also, without doubt, one of the paramount achievements of human culture. Buildings hold an infinite source of knowledge; they are historical documents in all their detail, regardless of whether they are original or later additions. We can always put new and as yet unasked questions about the past to them and they might have answers for us. Future generations may ask different questions and look for different answers. But this search for information works only if the informant actually is a witness of the times we are interested in, if it bears verifiable traces of the past. These traces – materiality, covered layers of lacquer, construction features, tool marks, inaccuracies and so on – disappear with the loss of the original. Buildings can undeniably be reproduced, but only as a result of our knowledge, not as a source for it. Their historicity is not (re)producible. Therefore, it seems misleading to designate them as monuments or heritage. Although their design features hark back to a historical basis, they are not inherited because the object meant to be handed down was destroyed by either nature or, worse, human failure. But these standards as set by the *Venice Charter* became subject to challenge before too long.

## POSTMODERN APPROACHES

Recent loss and destruction does not remain the only motivation for reconstruction. With the rise of the postmodern, the austere principles of modern architecture crumbled and gave way to a new devotion towards historicism. The Council of Europe declared 1975 as the European Architectural Heritage Year.

This accompanied the rediscovery of the historic city centres that were often neglected in the decades after the Second World War. The old buildings and quarters gained new attention after post-war urban planning revealed its deficits.[22] However, the new attention was less historical interest than affection for the mere picturesque qualities of historical work.[23] That same year, Umberto Eco, one of Europe's leading intellectuals, witnessed a change in people's attitude towards works of art in particular and towards reality in general. Travelling through the United States, he observed countless copies of artworks from European museums, as well as theme parks and J. Paul Getty's replica of a Roman villa. The perfection, the construction of the 'absolute fake' (Eco 31), strikes him, and the completeness of some works – despite the fragmentary state of their original – allow him to conclude, 'imitation has reached its apex and afterwards reality will always be inferior to it' (Eco 46). In this regard, Eco speaks about hyperreality, a condition that seems more obvious than reality itself. Concerning architectural reconstructions, the hyperreality refers to the denial of a building's history and its character as a historical document. For example, Berlin City Palace's brand new Baroque façades, coated around a ferroconcrete core, appear to erase the half-century of the building's complete absence – its replacement by the modernist Palace of the Republic, the seat of the parliament of the German Democratic Republic. With architecture's fundamental role for our understanding of history in mind, the reconstruction talks about an unbroken history; it creates a hyperreality that overlays real historic events.[24]

Truly striking in this regard is that along with a postmodern 'more casual attitude towards the problem of authenticity' (Eco 16), the attitude towards heritage in general has changed. This becomes obvious when we return to Ypres and Warsaw. What is of special interest is that both cities feature World Heritage sites. The Belfry of Ypres – completely reconstructed after the First World War – is mentioned in the UNESCO list as one of 32 belfries 'built between the 11th and 17th centuries' (UNESCO n.d.). This questionable evaluation and the inclusion of the tower raise the question of what actually defines historical architecture. Is its facsimile design enough to make it a historical monument? Is it warranted to disregard its true age, materiality and historical becoming? Ypres' inclusion in 1999 was ultimately justified by the inclusion of Warsaw in 1980 (ICOMOS 1998 137). The Historic Centre of Warsaw was then regarded as an 'exceptionally successful and identical reconstruction of a cultural property' (UNESCO 1980 4) and, thus, included into the World Heritage List. But it was included for the quality of its reconstruction, not for being an old and historic city. However, the inadequate evaluation that it be 'identical with the original' (ICOMOS 1980 1) paved the way for future inclusions of total reconstructions. Today, as a result, we find many (re)constructed buildings designated as architectural heritage.

Furthermore, reconstruction became an established method of urban planning. Recent years have witnessed a plethora of reconstructions of buildings that vanished decades or even centuries ago. Their absence is not accompanied by a sense of loss. Yet, huge efforts have been made to revive them. The Berlin City Palace is just one example, though a costly one right in the heart of a European capital.

I.2   Franklin Court:
Ghost Structure
by Robert Venturi
*Source:* Public
domain: http://
www.nps.gov/
storage/images/
inde/Webpages/
originals/377.jpg.

In other parts of the world, however, reconstruction has a historical background. Japan's famous Ise Shrines are an outstanding example of a structure that was completely rebuilt within a scheduled period of time. Every 20 years since the seventh century, the 123 buildings of the shrine have been replaced by identical reproductions made from freshly cut wood (Mager 99). Other historical buildings in Japan are also subject to dismantling and the replacement of their material in order to keep their appearance intact. Nevertheless, Japan's historical architecture has very distinctive features. It emerged from completely different cultural circumstances and is by no means comparable to Western stone buildings. Hence, it is unreasonable to cite Ise Shrines as a historical justification for Western reconstruction projects (Falser 86).[25] In effect, there are many ways of getting hold of the past and bygone works of architecture. When Robert Venturi was confronted with the task of reconstructing Franklin Court, the Philadelphia residence of Benjamin Franklin, due to the United States Bicentennial celebration in 1976, he provided an intelligent and stimulating design. Sticking to the facts, he only executed what was retraceable: the location and the building's dimensions (Figure I.2). By providing only the outlines, made of square steel tubes, he addressed the visitor's imagination and historical knowledge. Moreover, he did not determine the uncertainties concerning the appearance of the lost building but openly showed the haziness of the knowledge about it by providing leeway for different interpretations.[26]

A less radical strategy can be found in the recent renewal of Leipzig University, Germany's second-oldest university. The new Paulinum building, which contains the assembly hall and an oratory, was erected on the grounds of the thirteenth-century Paulinerkirche (University Church of St Paul). The church, a thorn in the eye of Walter Ulbricht – by then the East German head of state – was dynamited in 1968 to make way for a modern campus building. After the German reunification,

I.3  Leipzig: Paulinum, 2012
*Source:* Creative Commons (http://creativecommons.org/licenses/by-sa/3.0/deed.en):http://commons.wikimedia.org/wiki/File:Paulinum Augusteum72012.JPG.

discussion began to again replace the latter by a new construction. Plans to reconstruct Paulinerkirche caused a furore that eventually led to the withdrawal of the institution's Vice Chancellor, a vehement opponent to the reconstruction plans (Finger). The extensive dispute was finally arranged through a reasonable compromise suggested by Dutch architect Erick van Egeraat. He designed a modern building that provides reminiscence of the lost church by retracing its volume and featuring a Gothic tracery window and a rose window. Thus, the memory of the church could be revived and, simultaneously, a contemporary and spectacular building that fulfils the university's current needs was built (Figure I.3).

Considering these alternatives as well as contemporary potentials of construction and design, the question of why, at the end of the twentieth and the beginning of the twenty-first centuries, we suddenly rebuild so much from the past remains highly current. Does this imply a failure of contemporary architecture? Or is it simply an outcome of postmodern relativism? The following chapter case studies present original research that allow for an approximation of the questions raised by the topic. They show different examples of reconstructions, executed ones as well as merely planned ones and shed light on the motivations and expectations behind these projects. Apart from the meticulous reconstructions of buildings, they also deal with various approaches to get hold of vanished buildings, create ideal pseudo-historical quarters or adjust the image and identity of a city.

## CHAPTER CASE STUDIES

The chapter case studies in this book are arranged according to chronological aspects. Even though geographical jumps occur, this order allows for retracing the development from nineteenth-century restorations to full-scale reconstructions of bygone buildings and their designation as heritage in the late twentieth century. The first four chapters concentrate on the period until the middle of the twentieth century, whereas the subsequent chapters examine current projects. Apart from different settings in Europe, various places in Latin America, East Asia and the Middle East are covered whereby the worldwide emergence of the phenomenon is underlined.

In the first chapter, Arnold Bartetzky provides a historical overview of important nineteenth-century restoration projects in East Central Europe. He emphasizes the political ambitions accompanying the rediscovery of neglected or unfinished monuments. In times of emerging nationalism, they played an important role for the self-image and the territorial claims of the rival regional states. Therefore, their restorations and completions were equitable with the then contemporary attempts to restore and regain a sensed former glory of the respective empires. Bartetzky follows this until the emergence of post-war reconstructions of vanished buildings and cities. By shedding light on recent reconstructions from the post-Cold War era, political aims become apparent once again – this time to overcome the former Russian hegemony. By focusing on the projects presented, the author provides five theses to intellectually approximate the phenomenon. They include questions about the idealizing nature of reconstructions, their reasons, their political character and the willingness to officially accept them. The profoundness of the theses receives further support in the subsequent chapters since they can also be applied to the case studies represented therein.

Taking into account questions of authenticity and the creation of identity, Josep-Maria Garcia-Fuentes reveals the creation of Barcelona as a medieval city. Walking its streets today, Barcelona seems to be a gem of Gothic architectural ensembles. Few visitors and inhabitants know that they are strolling along a kind of stage when walking through the city. The famous Barri Gòtic (Gothic Quarter) is largely the outcome of a reinvention of the Catalan capital, realized between the 1920s and 1960s. It is remarkable that its artificial nature is hardly known and guides published after its completion focus on the Barri Gòtic as the remains of a glorious past. Garcia-Fuentes retraces the story of its construction in theory and practice. He explains the political ambitions and hypothetical constructs that lead to the architectural fantasy that eventually serves the demands and desires of tourism. The latter, in turn, arranges for the oblivion concerning the artificiality of the city centre. The case study allows for expanded comprehension of historical cities and heritage sites. Its appearance and 'historicity' are mainly outcomes of political and cultural discourse.

In Chapter 3, Robert Born takes a closer look at Communist Hungary and Romania. After the destructions of the Second World War, large-scale reconstruction projects were realized in both countries. In the case of the Royal Palace in Budapest, one of the largest excavations in Europe by the time was followed by a reconstruction of the palace including its bygone medieval features. Despite the fragmentary evidence, the palace was included on the UNESCO World Heritage List in 1987. The reconstruction of Trajan's victory monument in Adamclisi in Romania is a good example of a construction of history that links contemporary ideology with ancient monuments. It served Romanian leader Nicolae Ceauşescu as historical legitimation for his political agenda. Its continued use as a backdrop for political events even long after the end of Ceauşescu's regime underlines the broad scope of political claiming to which monuments are often subject.

The case of the Jesuit Church and College in São Paulo demonstrates the long process and the changes to which the memory of a building is subject. Renato Cymbalista and João Carlos Santos Kuhn examine the erection, abandonment and reconstruction of a building that throughout history played different roles for the identity of the Brazilian metropolis. By retracing the edifice's story under the aspects of toponymy, pictorial representation and lobbying, the authors point out how the half-century-long gap between the building's physical disappearance until its executed reconstruction is bridged in collective memory. Names and denominations are of equal importance as figurative visualization. Both are keeping the memory alive and are able to influence and manipulate that same memory. Vivid memory is able to raise the attention of a lobby that, in return, influences the characteristics of that memory to its own advantage. The desires and ideals created thereby eventually result in the reconstruction of a bygone building that now has a completely new connotation.

In Chapter 5, Julien Bastoen considers reconstructions as architectural cloning. He sheds light on three recent projects that aim to revive the memory of bygone French monarchic regimes. While only one, the Royal Gate of the Palace of Versailles, was actually carried out in 2008, the two others, the Tuileries Palace in Paris and Saint-Cloud Castle located between Paris and Versailles are analysed with regard to the propaganda campaign that initiated public awareness about the absence of the historic buildings. Being a project directly supported by the then right-wing government, the Royal Gate evokes an ideal eighteenth-century appearance of the palace. The examination of the privately initiated but eventually unrealized reconstructions of the Tuileries Palace and Saint-Cloud demonstrates broad public support. It underlines the popularity of historical sites of representation among wide circles of the general public as well as an increasingly 'more casual attitude towards the problem of authenticity' as Eco noticed it (Eco 1995 16).

Alice Y. Tseng deals with both diachronic and synchronic copying in East Asia. She explains two separate though ideologically similar cultural-historical enterprises in China and Japan that culminated in the same year, 2010. By taking a closer look at the imperial city Chang'an in China and Heijō Palace in Japan, she depicts the eighth-century modes of taking over the structure of an imperial city. Chang'an functioned as an antetype for Japanese and Korean urban planning. But the adoption of its plan is not the only similarity. Both sites were subject to decay from the tenth and ninth centuries respectively. After centuries of absence, they were rediscovered in the middle of the twentieth century. Excavations revealed that the wooden structures had decayed to mere imprints in the ground. However, the reclamation of antiquity was taken to the next step with the full-scale reconstruction of major structures. Despite poor evidence, the Chinese and the Japanese projects received direct support from UNESCO. The results, Daming Palace Heritage Park and the reconstructed Heijō Palace, are two thrilling examples of hypothetical reconstructions that rest upon scientific research but almost completely lack certainty. In both cases, the authentic heritage site was turned into a kind of theme park, illustrating contemporary knowledge about the buildings rather than allowing for an examination of the remains.

In Chapter 7, Jing Zhuge investigates the motivations behind the project of reconstructing the fifteenth-century Great BaoEn Pagoda in Nanjing, also known as Porcelain Tower. The building, one of China's tallest pagodas, was destroyed during the Taiping Rebellion in the 1850s. However, it was handed down in illustrations from European travellers and survived partially in the form of relics used to erect other buildings. Scientific interest in the building emerged in the 1980s; by 2001, the Nanjing Municipal Government initiated a reconstruction project. By analysing media contributions, the author reveals the character of the city's interest in revitalizing the building. Formerly added to the Seven Wonders of the World by famous seventeenth-century traveller and China expert Johan Nieuhof, it not only played an important role in the city's local identity but also contributed to its international fame. Businessmen and local politicians quickly discovered the commercial power of the spectacular project. Thus, the reconstruction is largely seen as a business venture aiming at economic benefit.

The final chapter by Alexandra Klei provides an insight into Tel Aviv's strategies to cope with its image as the 'White City', derived from its Bauhaus heritage. Being recognized as a UNESCO World Heritage Site (White City of Tel-Aviv – the Modern Movement) in 2003 contributed to the city's identity as a centre of modernist architecture. However, a closer look at the designated areas revealed little unity in the stylistic appearance of the existing buildings. Therefore, the authorities were forced to (re)construct street pictures in keeping with the stated heritage status, to adjust existing architecture to the image. Reflecting on the relationships between architecture, memory and reconstruction, Klei addresses basic problems related to the retroactive idealization of historical buildings. Even though the reconstructions executed in Tel Aviv are rather small in scale, the chapter examines the process as an effective means to construct a smooth yet unilateral image of history.

Although the case studies represent only a clipping of worldwide reconstruction projects, they allow for a deeper understanding of the phenomenon. While focusing on different aspects of the motivations for the reestablishment of lost buildings, they provide useful answers on fundamental questions concerning the topic. They reveal the recurring motive of glorified national histories, the representation of power in connection with historical legitimation, the compensation for loss and suffering, the visually retraceable building of identity and the fondness for the beauty of historical forms. Yet, they are primarily thought to stimulate further investigation and to ask further questions about the meaning of architecture, the dubious persuasiveness of contemporary architecture, our relation with the past and the ways to get hold of it. The critical questioning of reconstruction projects within this book is not aimed at their condemnation or at the denial of their right to exist. It is aimed at the formation of a prolific dialogue. The current wave of reconstruction is an expansive and complex topic that reveals much about our time and the contemporary view of the world. Simply denying it as mere nostalgia or even self-deception fails to cope with the reasons for its popularity. Reconstruction can help to heal deep wounds as it did in Warsaw and Dresden.[27] Notwithstanding this, we should be very careful about our heritage – about the things that really witnessed the past, the way we treat it, what we can learn from it and its loss, and our authentic application in today's world and its architecture.

## BIBLIOGRAPHY

Baudrillard, Jean. *Simulacres et Simulation*. Paris: Éditions Galilée, 1981. Print.

Baus, Ursula and Michael Braum, eds. *Rekonstruktion in Deutschland. Positionen zu einem umstrittenen Thema*. Basel: Birkhäuser, 2009. Print.

Buttlar, Adrian von, Gabi Dolff-Bonekämper, Michael S. Falser, Achim Hubel and Georg Mörsch. *Denkmalpflege statt Attrappenkult: Gegen die Rekonstruktion von Baudenkmälern – eine Anthologie*. Basel: Birkhäuser, 2010. Print.

Choay, Françoise. *The Invention of the Historic Monument*. Cambridge: University Press, 2001. Print.

Dehio, Georg. 'Denkmalschutz und Denkmalpflege im Neunzehnten Jahrhundert' (1905). *Denkmalpflege. Deutsche Texte aus Drei Jahrhunderten*. Ed. Norbert Huse. Munich: Beck, 2006. 139–46. Print.

De Naeyer, André. 'La Reconstruction des Monuments et des Sites en Belgique Après la Première Guerre Mondiale'. *Monumentum*, 20–22 (1982): 167–88. Print.

Derrida, Jacques. 'Letter to Peter Eisenman'. *Art and Its Significance: An Anthology of Aesthetic Theory*. Ed. Stephen David Ross. Albany: Suny Press, 1994. 429–38. Print.

Eco, Umberto. *Faith in Fakes: Travels in Hyperreality*. London: Minerva, 1995. Print.

Falser, Michael S. *Zwischen Identität und Authentizität: zur Politischen Geschichte der Denkmalpflege in Deutschland*. Dresden: Thelem, 2008. Print.

——. 'Von der Charta von Venedig 1964 zum Nara Document on Authenticity 1994: 30 Jahre 'Authentizität' im Namen des Kulturellen Erbes der Welt'. *Renaissance der Authentizität? Über die neue Sehnsucht nach dem Ursprünglichen*. Eds. Michael Rössner and Heidemarie Uhl. Bielefeld: Transcript, 2012. 63–87. Print.

Fenlon, Iain. *Piazza San Marco*. Cambridge: Harvard University Press, 2012. Print.

Finger, Evelyn. 'Die Angst vor der Kirche'. *Zeit*. 30 May 2008. Web. 10 January 2015. <http://www.zeit.de/2008/23/Leipziger-Bilderstreit>.

Fischer, Manfred F. 'Campanile von San Marco, Venedig, Italien'. *Geschichte der Rekonstruktion – Konstruktion der Geschichte*. Eds. Winfried Nerdinger, Markus Eisen and Hilde Strobl. Munich: Prestel, 2010. 342–4. Print.

Glendinning, Miles. *The Conservation Movement: A History of Architectural Preservation: Antiquity to Modernity*. London: Routledge, 2013. Print.

Goethe, Johann Wolfgang von. 'Von deutscher Baukunst' (1771). *Denkmalpflege. Deutsche Texte aus drei Jahrhunderten*. Ed. Norbert Huse. Munich: Beck, 2006. 23–5. Print.

Hassler, Uta. *Das Prinzip Rekonstruktion* [Proceedings from the conference 'Das Prinzip Rekonstruktion'], Institut für Denkmalpflege und Bauforschung (IDB), ETH Zürich. 24–5 January 2008]. Zurich: Hochschulverlag der ETH, 2010. Print.

ICOMOS. *Il Monumento per L'Uomo: Atti del II Congresso Internazionale del Restauro Venezia, 25–31 Maggio 1964* [The Monument for the Man: Records of the II International Congress of Restoration]. Padua: Marsilio, 1971. Print.

ICOMOS. 'Advisory Body Evaluation – The Historic Center of Warsaw'. 1980. Web. 31 July 2014. <http:// whc.unesco.org/archive/advisory_body_evaluation/030.pdf>.

ICOMOS. 'Advisory Body Evaluation – Flemish Belfries'. 1998. Web. 20 July 2014. <http://whc. unesco.org/archive/ advisory_body_evaluation/943bis.pdf>.

ICOMOS. 'Introducing ICOMOS'. 2011. Web. 11 February 2014. <http://www.icomos.org/en/ about-icomos/mission-and- vision/mission-and-vision>.

Jencks, Charles. *The Language of Post-modern Architecture*. London: Academy Editions, 1991. Print.

Jokilehto, Jukka. *A History of Architectural Conservation*. Oxford: Butterworth-Heinemann, 1999. Print.

Mager, Tino. *Schillernde Unschärfe. Der Begriff der Authentizität im architektonischen Erbe*. Dissertation. Berlin Institute of Technology, 2014.

——. 'Die Umhüllung der Ewigkeit: Japans Ise Schreine – Meisterwerke des Verborgenen'. *Haut und Hülle – Umschlag und Verpackung*. Eds. Michael Fisch and Ute Seiderer. Berlin: Rotbuch 2014b. 90–101. Print.

Meissner, Irene. 'Filarete-Turm des Castello Sforzesco, Mailand, Italien'. *Geschichte der Rekonstruktion – Konstruktion der Geschichte*. Eds. Winfried Nerdinger, Markus Eisen, and Hilde Strobl. Munich: Prestel, 2010, 302–4. Print.

Milner, John. *A Dissertation on the Modern Style of Altering Ancient Cathedrals, as Exemplified in the Cathedral of Salisbury*. London: J. Nichols, 1798. Print.

Nerdinger, Winfried, Markus Eisen and Hilde Strobl, eds. *Geschichte der Rekonstruktion – Konstruktion der Geschichte*. Munich: Prestel, 2010. Print.

Nietzsche, Friedrich. *Vom Nutzen und Nachtheil der Historie für das Leben*. Munich: dtv, 1996. Print.

Omilanowska, Małgorzata. 'Rekonstruktion Statt Original – das Historische Zentrum von Warschau'. *Informationen zur Raumentwicklung*, 3.4 (2011): 227–36. Print.

Pfarre-St.-Josef-ob-der-Laimgrube. 'Geschichte im Detail'. n.d. Web. 3 December 2012. <http://www.pfarrelaimgrube.at/stjosef/index. php?mid=Kultur&cid=Geschichte>.

Pittarello, Silvia. '1902–2002. Centenario del Crollo del Campanile di San Marco. "Morte e miracoli del pandolone muto"'. 2002. Web. 11 February 2014. <http://www. culturaspettacolovenezia.it/?iddoc=7218&page=3>.

Rettig, Manfred, ed. *Rekonstruktion am Beispiel Berliner Schloss aus Kunsthistorischer Sicht: Ergebnisse der Fachtagung im April 2010 – Essays und Thesen.* Stuttgart: Steiner, 2011. Print.

Ruskin, John. *The Seven Lamps of Architecture.* Orpington: George Allen, 1889. Print.

Schediwy, Robert. *Rekonstruktion: Wiedergewonnenes Erbe oder Nutzloser Kitsch?* Münster: LIT, 2011. Print.

Schneider, Helmut. 'Nichts als sichtbare Geschichte?' *Zeit Online.* 1975. Web. 18 June 2013. <http://www.zeit.de/1975/32/nichts- als-sichtbare-geschichte>.

Storer, Henry Sargant and James Sargant Storer. *History and Antiquities of the Cathedral Churches of Great Britain: Salisbury, Gloucester, Hereford, Chester, Worcester, Lichfield, Carlisle.* London: Rivington, 1816. Print.

Thomson, Robert Garland. 'Authenticity and the Post-conflict Reconstruction of Historic Sites'. *CRM Journal* 5.1 (2008): 64–80. Print.

Tschudi-Madsen, Stefan. *Restoration and Anti-Restoration: A Study in English Restoration Philosophy.* Oslo: Universitetsforlaget, 1976. Print.

UNESCO. 'Belfries of Belgium and France'. n.d. Web. 31 July 2014. <http://whc.unesco.org/en/list/943/>.

UNESCO. 'World Heritage Committee. 4th Session (Paris, 19–22 May 1980) Report of the Rapporteur'. 1980. Web. 31 July 2014. <http://whc.unesco.org/archive/repbur80.htm>.

Viollet-le-Duc, Eugène-Emmanuel. *Dictionnaire raisonné de l'architecture française du XIe au XVIe siècle: Tome 8 [Qua – Syn].* Paris: Morel, 1866. Print.

Wörner, Ernst, ed. 'Generalversammlung des Gesammtvereins der Deutschen Geschichts- und Alterthumsvereine zu Kassel'. *Correspondenzblatt des Gesammtvereins der Deutschen Geschichts- und Alterthumsvereine* 30.11 (1882): 81–3. Print.

## NOTES

1   The Cathedral of Christ the Saviour was consecrated in 1883, destroyed by dynamite under Stalin in 1931, replaced by an open-air swimming pool, and eventually reconstructed between 1994 and 2000. The construction of the Berlin City Palace started in the fifteenth century, was altered several times, damaged in the Second World War and demolished in 1950. The Palace of the Republic, which is the seat of the parliament of the German Democratic Republic, was erected on the site. This was demolished in 2008 to allow for the reconstruction of the Berlin City Palace from 2013 onwards. Gyeongbok Palace was completed in 1395, destroyed in 1592, rebuilt in 1868 and again destroyed under Japanese rule. It has been under reconstruction since 1990.

2   For example the Palace of the Grand Dukes of Lithuania, Daming Palace near Xi'an or Heijō Palace in Nara; see Chapters 1 and 6.

3   While much debate on recent reconstruction projects in Germany (Berlin City Palace, Potsdam City Palace, Dresden's Altmarkt area, Brunswick Palace and Paulinerkirche) have taken place and several publications are available (Baus, Buttlar, Falser, Hassler, Nerdinger, Rettig, Schediwy), the topic is poorly covered in international specialist literature.

4   For a comprehensive history of architectural heritage and conservation/restoration, see Choay and Jokilehto.

5    For example the Porta Nigra in Trier, the Colosseum in Rome, the Heidelberg Castle, the Cologne Cathedral and Saint-Ouen in Rouen.

6    The controversially received exhibition *Geschichte der Rekonstruktion – Konstruktion der Geschichte* (Architekturmuseum der TU München, 22 July–31 October 2010) proved that there are no cases of meticulous reconstructions of lost buildings before the twentieth century. The numerous examples of rebuilding and completion rarely show an interest in precisely regaining the visual features of the lost building (Nerdinger).

7    See Nietzsche.

8    For the nineteenth-century development of the restoration movement, see Tschudi-Madsen.

9    In particular, Johann Wolfgang von Goethe's essay 'Von deutscher Baukunst' (1771), an anthem to the Strasbourg Cathedral and Gothic architecture, led to a novel evaluation of medieval buildings in the German countries.

10   Friedrich Nietzsche heavily criticizes the uncritical embracing of the past that he witnessed among his fellows in *Vom Nutzen und Nachtheil der Historie für das Leben* (Nietzsche).

11   'rétablir dans un état complet qui peut n'avoir jamais existé à un moment donné' (translation by author). However, Viollet-le-Duc's approach must not be underestimated. He possessed keen knowledge about medieval architecture and believed he was able to reconstruct medieval buildings due to the inherent logic of Gothic architecture. It was his aim to continue it and to bring it to perfection. This can also be seen in his designs that combine Gothic elements with contemporary iron structures.

12   For an overview see Falser (2008 43–70).

13   The term goes back to archivist Hermann Grotefend, who proclaimed it in 1882 in correlation with a resolution regarding Heidelberg Castle (Wörner 81).

14   *Law for the Preservation of Old Shrines and Temples* (Koshaji Hozonhō), 10 June 1897.

15   Many travellers expressed their feelings of grief when they saw St Mark's Square without the Campanile (Fischer 343).

16   Even during Renaissance, the fascination for classical works only generated reconstructions on paper – for example Juan Bautista Villalpando's reconstruction of Solomon's Temple (1604) or Raffael's plan *Roma Instauranda.*

17   In 1883, Luca Beltrami, the executing architect, had already researched the possibility of reconstructing Sforzesco Castle's Filarete Tower in Milan. The fifteenth-century tower was destroyed in 1521 and eventually reconstructed between 1903 and 1905 (Meissner 303).

18   'sarebbe un voler falsificare la storia dell'architettura se si ricostruisse il campanile nello stile antico' (translation by author).

19   However, the charter accepts anastylosis as a reconstruction technique for archaeological sites.

20   ICOMOS's work is based on the *Venice Charter* (ICOMOS 2011). In the *World Heritage Convention*, ICOMOS is recognized as the authority concerning conservation and preservation (ICOMOS 1971 LXXII–LXXIV).

21   For example the *Declaration of Dresden on the 'Reconstruction of Monuments Destroyed by War'* (1982), *ICOMOS Krakow Charter* (2000) or *Riga Charter* (2000).

22   The most ostensible example of planning failure is the demolition of Pruitt–Igoe, which started on 16 March 1972. Architecture theorist Charles Jencks called that event 'the day that modern architecture died' (Jencks 23).

23   For a revealing example of historical awareness at the time, see Schneider.

24   A few years after Eco, Jean Baudrillard also deals with hyperreality. In *Simulacres et Simulation*, he writes about the effects of the increasing replacement of real things by copies and simulacra, and the subsequent loss of the real.

25   For a comprehensive investigation, see Mager 2014b.

26   There exist only a few comparable examples, among them Juan Garaizabal's outlines of Berlin's Bethlehemskirche or Cologne's Auferstehungskirche by Frantisek Sedlacek.

27   I am referring to the reconstruction of Warsaw's historic centre and to the reconstruction of Dresden's Frauenkirche.

# 1

# Architecture Makes History: Reconstruction and Nation-building in East Central Europe

*Arnold Bartetzky*

Since the 1980s, we have observed the emergence of a new commitment to restoring destroyed buildings and urban structures in various parts of Europe. The history of architectural reconstruction, however, is much longer. From the Romantic era onwards, the nineteenth and twentieth centuries saw several waves of more or less historically accurate reconstruction projects. In many cases, these activities were closely associated with political attempts to glorify the past for present-day purposes. Particularly in the eastern part of Europe, the reconstruction of monuments heavily loaded with symbolism has played a significant role in the processes of nation-building as well as reawakening and redefining national identities. This specific notion of reconstruction was triggered by the historical experiences of the countries between Germany and Russia in the nineteenth and twentieth centuries, such as overdue national emancipation after long periods of foreign rule, several regime changes, the devastation of the two world wars, the emergence of new and the rebirth of old states, the frequent shift of borders and population displacement. Ethnic tensions and political strife were often accompanied by the symbolically-motivated destruction of significant monuments, which were perceived by the communities affected as attacks on their national identity.

Accordingly, national self-assertion by reviving destroyed architectural heritage has been and remains one of the central tasks of reconstruction projects in this part of Europe. This paper explores this aspect of reconstruction politics by referring to selected examples ranging from the Prussian revival of the Teutonic Knights' Marienburg Castle (Malbork) in the nineteenth century to Polish projects before and after 1945 and present-day reconstruction campaigns in post-Soviet states.[1]

## RESTORATIONS AND RECONSTRUCTIONS IN THE NINETEENTH AND EARLY TWENTIETH CENTURY

Since the nineteenth century, the reconstruction of destroyed buildings of symbolic value as well as the reconstructive restoration of substantially-altered edifices and the completion of unfinished ones have been seen as important to nation-building and national self-assertion. Although this nationalist element has never been the only reason for tackling schemes of this nature, it has frequently been the driving force behind them in newly emerged nation states and after fundamental political changes. The celebration of historical architecture has been used as a means to enhance national identity, and sometimes to legitimize state sovereignty or even express territorial claims. This national function of reconstruction can be explored by studying some examples in central and eastern Europe from the nineteenth century and from the major turning points of 1918, 1945 and 1989.

One of the earliest German examples of restoration in the service of the nation was Marienburg (now known as Malbork) Castle (Figure 1.1). The rebuilding of Europe's biggest red brick castle, which had become badly dilapidated and been substantially altered over the centuries, began in 1817, and took more than 100 years to complete. The initially regional nature of this project recalling the medieval roots of the kingdom of Prussia gave way over time to an increasingly pan-German, aggressive character. After the foundation of the German Empire in 1871, the former centre of power of the Teutonic Order was ever more interpreted as a stronghold of Germanness in the formerly Polish eastern territories, which had been annexed by Prussia in late eighteenth century and were now characterized by national tensions between the German and Polish populations. Accordingly, German historian Heinrich von Treitschke praised the reconstruction of Marienburg as 'a monument of victory for the old State of the Teutonic Order, which was so proud to have awakened the other Germans to their holy struggle'[2] (Boockmann 164).

Strengthening national identity by remembering the commonly idealized period of the German Middle Ages was also the aim of restoring Wartburg Castle in Thuringia. Although the project was less aggressive in nature than the rebuilding of Marienburg, it was no less symbolic as the decaying building was associated with Martin Luther and the Reformation as well as the heyday of the medieval German Minnesang (courtly love song) and the benevolent work of St Elisabeth of Thuringia, one of the most popular German saints. Grand Duke Carl Alexander von Sachsen-Weimar-Eisenach, the instigator of the project, intended the rebuilt Wartburg Castle to become an inspirational national monument to German culture. With this in mind, the painter Carl Alexander Simon praised him thus: 'You will be the first … to open up to the nation the shrine in which it will find the documents of its glory … and the erection of the Wartburg will mark the beginning of the magnificent epoch of German self-awareness'[3] (Krauß 15).

Both projects were more or less clearly perceived by their contemporaries as a symbol of national awakening. This is even truer of the resumption of the building of Cologne Cathedral, which had remained a torso since sixteenth century. The reconstruction work, which began in 1842, fuelled hopes for the emergence

1.1   Marienburg, Upper Castle, East side after reconstruction, photograph around 1900 *Source:* Public domain: http://commons.wikimedia.org/wiki/File:Marienburg 19001.jpg.

of a new German empire. Back in 1814, writer and journalist Joseph Görres had appealed for 'Germany's forces to be united to complete the Gothic cathedral'. He saw the centuries-long suspension of building work as a metaphor for the 'shame and humiliation' of Germany, a nation which had fallen victim to 'its own discord and foreign impertinence'. He therefore called for the completion of the cathedral as 'a symbol of a new empire which we now want to build'[4] (Görres 5–7).

Another equally ambitious yet unfinished building project dating back to the Middle Ages was St Vitus's Cathedral in Hradčany, the age-old centre of former Bohemian statehood in Prague. Construction work resumed in the 1860s following (as in Cologne) lengthy debate and planning. The completion of the project was seen as a mark of veneration for the prosperous medieval heritage of Bohemia, the cradle of Czech statehood, and it is no coincidence that it took place at a time when the country was striving for political autonomy within the Austro-Hungarian Empire. Marie Kostílková, the author of a Czech publication on St Vitus's Cathedral, put it as follows: 'In a time when Prague had become an insignificant provincial city, and the city's cultural and social life was oppressed by a rigid centralism and the policy of Germanization, the unfinished cathedral recalled the glorious past of the Bohemian Kingdom ... and it aroused the idea to complete it'[5] (Kostílková 13). When the finished cathedral was consecrated in 1929, Czech hopes for national self-determination had already been fulfilled by the foundation of an independent Czechoslovak state.

For the Poles, Wawel in Kraków was just as significant to the Poles as Hradčany was to the Czechs. The former royal castle on Wawel Hill, a masterpiece of Renaissance architecture, was regarded by Poles as an architectural symbol of their political independence stretching back to the early Middle Ages. However, it was in a very poor condition, having been burned down at the beginning of the eighteenth century, rebuilt in a utilitarian manner, and turned into a barracks for the Austrian army shortly after the Third Polish Partition in 1795, depriving the Poles of their own statehood. The most striking sign of its decline was the dilapidated appearance of what had once been a highly elegant arcade yard, which had been walled up when used as a barracks. In the increasingly liberal atmosphere under Austria's Emperor Franz Josef, in 1905 the semi-autonomous local Polish administration recovered Wawel Castle from the Austrian military. There soon followed a campaign for its extensive, reconstructive restoration. The project's initial aim was to restore the castle's appearance as it had been in Poland's golden age and hence uplift the nation in the absence of an independent state. Construction work began under Austrian rule, and its completion after the First World War and the subsequent re-establishment of the Polish state were celebrated with a new self-assurance as the rebirth of a 'monument of the greatness and culture of the nation'[6] (Dettloff, Fabiański and Fischinger 75).

In the former capital Kraków and also other parts of interwar Poland, several architectural symbols of national culture which had suffered under foreign rule after the Polish Partitions were also restored to more or less their original form. One prominent example is the neo-classical Staszic Palace in Warsaw, the former seat of the influential Polish Society of Friends of Learning. In the 1890s it was converted into a Russian Orthodox church in an Old Russian, neo-Byzantine style. For the Poles, this callous, highly conspicuous example of Russification was a symbol of the suppression of their national culture by the occupying power. Soon after Poland's rebirth in 1918, it was rolled back by restoring the building's neo-classical form. For its supporters, reconstruction was nothing less than 'a dictate of national conscience'[7] (Dettloff 230).

## WAVES OF RECONSTRUCTION PROJECTS AFTER 1945 AND 1989

The widespread practice of reconstruction in the service of the nation in interwar Poland paved the way for the country's major reconstruction campaigns after the Second World War. Particularly the rebuilding of the old town in Warsaw was seen as a sign of resilience and as an act of cultural self-defence by a nation whose very survival had been threatened by the German occupation. For the new Communist regime, which had been installed by the Soviet Union and was hence largely perceived as an agent of a new occupying power, the reconstruction campaign was a welcome opportunity to present itself as a custodian of national heritage.

Far less well known than the large-scale rebuilding projects in Warsaw and other historical urban centres in Poland is the rebuilding of the Bethlehem Chapel in Prague (Figure 1.2). It was carried out at roughly the same time, a few years after the Second World War, as an extraordinary example of reconstruction in the

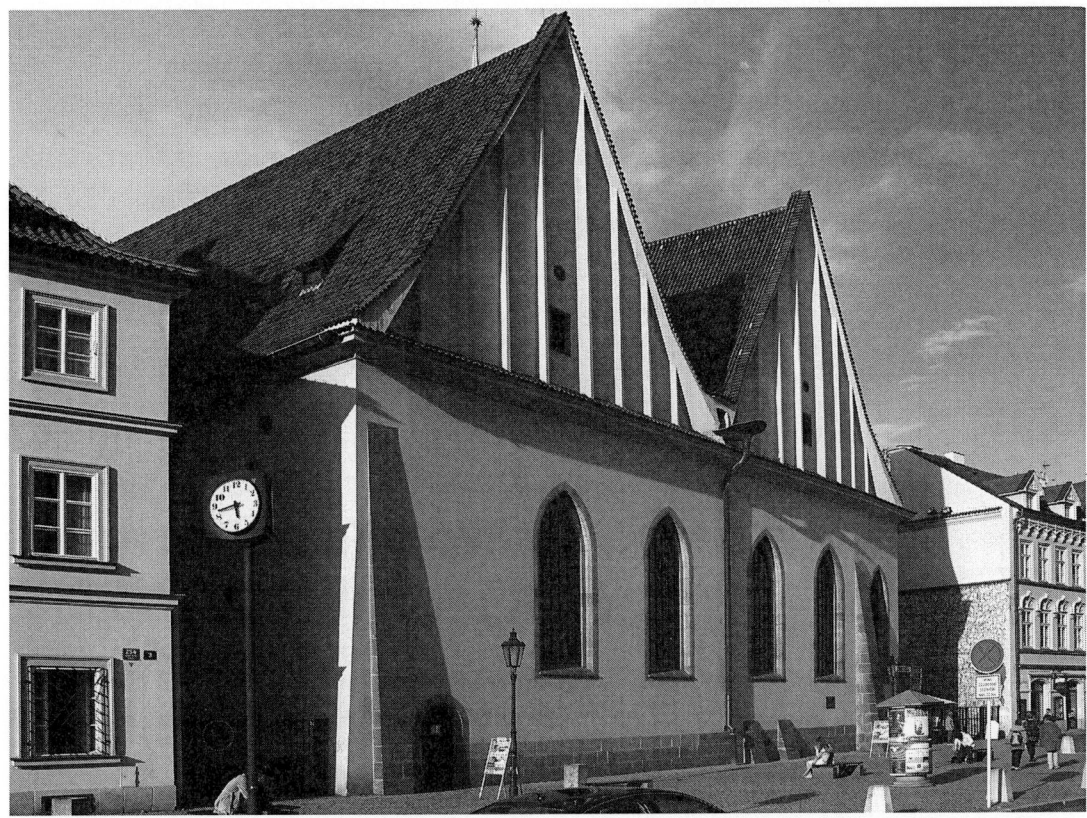

1.2   Prague, Bethlehem Chapel, photograph 2006 *Source:* Public domain: http://commons.wikimedia.org/wiki/File:Betlemska_kaple.jpg.

service of a narrative of national history. The original building had been erected in the late fourteenth century. In contrast to the majority of the churches in the city, which at that time was dominated by a German-speaking population, sermons in the Bethlehem Chapel were preached in Czech. Its most prominent preacher in the early fifteenth century was the Czech reformer Jan Hus. In the seventeenth century, during the Counter-Reformation, the chapel was taken over by the Jesuit Order. In the late eighteenth century, it was almost completely demolished: only the exterior walls remained and were later incorporated into a residential building. In 1950–52, more than one and a half centuries after its demolition, the chapel was rebuilt to a plain design on the orders of the new Communist Czechoslovak government, resulting in a vague approximation of the original. The main reason for its reconstruction was the building's symbolic significance for Czech national history in general, and for the Czechs' alleged perpetual struggle against German domination in particular. Jan Hus was regarded by Czech historiography as an eminent precursor of the national liberation movement. Accordingly, writer Josef Kajetán Tyl praised the chapel in the nineteenth century (when it did not exist) as a 'Bethlehem from which the Czech Messiah will arise'[8] (Nerdinger 282). A few years after the Second World War, which had ended in Czechoslovakia with the expulsion of the German population, the re-erection of the chapel was meant not only as an act of reverence towards the national hero Jan Hus, but at least implicitly also as an anti-German gesture of triumph. Moreover, Hus was adopted by Communist historiography as a social revolutionary and a progenitor of the new regime. According to a book published shortly after the reconstruction, he had turned the chapel into the 'the first people's tribune in the world', thus making it 'the origin of the Hussite movement, the major revolutionary stage in Czech history'[9] (Kubiček 111).

Since 1989, the formerly socialist part of Europe has seen a fresh proliferation of reconstruction projects. Reconstruction has become an important means of symbolic politics again in the service of national identity, especially (but not only) in the successor states of the Soviet Union.

One outstanding case is the rebuilding of the Cathedral of Christ the Saviour in Moscow in 1994–8. The original building, an outstanding example of neo-Byzantine Russian eclecticism, was inaugurated in 1883 as the foremost Orthodox church in the Russian Empire and as a memorial church commemorating Russia's victory over Napoleon. Therefore, as well as being the most important place of worship for the Orthodox Russian Church, it was also a highly prestigious edifice of the tsarist empire. In 1931, it was demolished on Stalin's orders to make way for the planned Palace of the Soviets, which was to be the tallest building in the world. However, construction work only got as far as the foundations. Once the project had been abandoned in the 1950s, a huge swimming pool was built on the site. In the meantime, public memory of the former cathedral gradually faded. In fact several decades after its destruction, few Muscovites still knew what it had looked like because all photographs showing the cathedral were banned. This all changed during perestroika. After the dissolution of the Soviet Union, rebuilding the cathedral even became an official project of the Russian state. Nowadays, the gold on the reconstructed cathedral appears to gleam more brightly than on the original. It symbolizes the restoration of the Christian religion in Russia and the vindication of the tradition of the tsarist Russian Empire with its unity of church and state. Moreover, by commemorating the victory over Napoleon, it also visualizes the continuity of Russia's claim to be a great power in world politics. One of its major supporters, Moscow's former mayor Yuri Luzhkov, praised the cathedral as a 'unifying symbol of the rebirth of the Russian nation' (Akinsha and Kozlov 156).

In the non-Russian territories of the former Soviet Union, however, several reconstruction projects symbolize not only national rebirth after overcoming the Communist regime, but also their efforts to assert themselves against Russia, their mighty neighbour. This dual function is eminently performed by the reconstruction of St Michael's Monastery in the Ukrainian capital Kiev, clearly inspired by the rebuilding of the cathedral in Moscow. The parallels in the history of the destruction of the two buildings as well as in the genesis of their reconstruction are indeed striking.

The monastery in Kiev, which was founded in the twelfth century by Prince Sviatopolk II and later extended in the style of the 'Ukrainian Cossack Baroque', was regarded as a place of eminent significance for the pre-modern tradition of the Ukrainian national state. Being a highly symbolic monument of Christianity

1.3    Vilnius, Palace of the Lithuanian Grand Dukes, photograph 2012 *Source:* Arnold Bartetzky.

as well as of the Ukrainian nation's striving for independence, the monastery was demolished in 1936 on the orders of the Stalinist Soviet Ukrainian government. Although the regime planned to erect a government complex on the site, only the adjacent building of the present Ministry of Foreign Affairs was built. The destruction of the national monument signified the Soviet oppression of Ukraine: in Ukrainian collective memory, it symbolized the repression of their national culture as well as the political mass murders of the Stalin era. The reconstruction of the cathedral in 1997–2000 expressed the nation's moves to distance itself from its Soviet past, the overcoming of Communism, and the reinstatement of an independent Ukrainian national history.

Another symbol of overcoming Russian hegemony is the reconstruction of the Palace of the Lithuanian Grand Dukes in Vilnius in 2002–13 (Figure 1.3). The palace, which was medieval in origin and revamped in the Renaissance period, had been destroyed not by the Soviet Union but by its predecessor, the equally imperialistic tsarist Russian Empire. The ramshackle building, which had been gradually decaying since the seventeenth century, was torn down a few years after the Third Polish Partition in 1795, when the Lithuanian part of the former Polish-Lithuanian Union was annexed by Russia. The palace's reconstruction was meant to emphasize the longevity of the Lithuanian state and hence historically justify its newly regained independence. Consequently, it has been promoted as 'the recovery of a symbol important for national identity and for historical commemoration' and even 'the expression of Lithuania's state sovereignty'[10] (Hinterkeuser 181).

## RECURRING PATTERNS

The buildings mentioned present only a few examples of numerous reconstruction projects which have been explicitly motivated by the political needs of a nation. Although this list of nationally charged rebuilding campaigns could be continued, let us turn to some common patterns of the development, execution and propagation of such projects. These patterns are explained below by five theses along with corresponding examples.

1.    *Architectural reconstruction is always based on a selective relationship with the original. The point of reference is usually not the heterogeneous state of the building at the moment of its destruction, but an idealized, stylistically homogeneous form from an earlier period regarded as the golden age of the nation or the city.*

Particularly vivid examples of this selective procedure are the reconstruction practices in Polish historical city centres such as Warsaw, Gdańsk and Poznań after the Second World War. During these rebuilding campaigns, numerous preserved remnants of nineteenth-century eclecticism and early modernism were removed or redeveloped in favour of reconstructed or vaguely imitated forms from earlier periods. In this way, reconstruction was used as an opportunity to stylistically and historically purify the cityscape. In several cases, this even involved the demolition of completely preserved buildings. In Gdańsk and Poznań, which had been under Prussian rule since the Second Polish Partition of 1793, the disdain of architecture of the nineteenth and early twentieth centuries, which was a common phenomenon throughout Europe at that time, was enhanced by the political aim to supress the memory of the German periods of the cities' history. An exemplary case of this architectural Polonization is the former patrician houses on Poznań's market square. The architectural variety from the time before the Second World War was widely replaced by a homogeneous vision of a basically Renaissance-styled old Polish city which had never existed in this form. This observation is closely linked to the second thesis.

2.    *Idealized conceptions of the past often triumph over historical accuracy. Consequently, many reconstruction projects have been carried out despite poor documentation of the original building.*

Particularly in the nineteenth and early twentieth centuries, this was common practice. Extensive parts of Marienburg and Wartburg as well as the interiors of Wawel Castle were rebuilt without a precise knowledge of the buildings' original appearance. After the Second World War, in the large-scale reconstruction campaigns in Poland, 'correcting the past' by deliberately diverging from the original became the rule. And even nowadays, planners of reconstruction schemes sometimes resort to using their imagination. For example, in the Palace of the Grand Dukes in Vilnius, some 'historical' interiors were created despite the lack of either pictorial sources or detailed descriptions documenting the rooms' former appearance. To complete the historical appearance, furnishings were procured on the international antiques market.

The palace in Vilnius also provides an example for the following thesis.

3.    *The circumstances of the destruction of a building determine its future likelihood of being reconstructed. This probability is especially high if the destruction was carried out deliberately by a foreign power, as an attack on the national culture and identity, or at least if those calling for reconstruction can insinuate such intent. In such cases, reconstruction is welcomed as an act of overcoming national humiliation.*

For example, for Poland's conservator general of historical monuments Jan Zachwatowicz, the reconstruction of the old town of Warsaw, which had been deliberately and systematically ravaged by the German occupiers, was imperative not just because of the unprecedented scale of destruction but especially because 'several historic buildings [had] been destroyed not by accident, in armed struggles, but as a deliberate act by the Nazis to eradicate Polish cultural achievements'[11] (Zachwatowicz 1965 44). Speaking in the heroic tone of his time, he pledged: 'We will not accept the annihilation of our cultural monuments. We shall reconstruct them, we shall rebuild them from their foundations, in order to hand over to later generations if not the authentic, at least the precise form of these monuments as it is alive in our memory ... '[12] (Zachwatowicz 1946 48).

Several historical buildings which had been destroyed by German troops were also rebuilt in the post-war Soviet Union, including the residences of the tsars in Peterhof and Tsarskoye Selo near St Petersburg as well as much of the old town and even several churches in Novgorod. This reconstruction work was ordered by the central Soviet government and began even before the Second World War had ended. This is astonishing, for the Stalinist Soviet Union had little appreciation for tsarist palaces and had demolished churches on an unprecedented scale. One of the reasons for the special treatment of monuments in St Petersburg and Novgorod was the circumstances and perpetrators of their destruction. A Soviet travel guide for Novgorod states:'During the Great Patriotic War of 1941–5, the fascist barbarians intended to level Novgorod to the ground. ... The Special State Commission for the Investigation of the Atrocities of the Fascist Occupiers assessed the damage to the city and found it exceeded a billion roubles'.[13] The immediate reconstruction of the destroyed monuments was seen as an effective propaganda means to demonstrate the Soviet Union's invincibility and its triumph over Nazi Germany.

Half a century later, post-Soviet states decided to reconstruct numerous buildings demolished for symbolic reasons in the Soviet era. The re-erection of St Michael's Monastery in Kiev for instance is meant to roll back a Soviet act of destruction which was perceived (almost certainly not inaccurately) as a brutal attack on the Ukrainian church and Ukraine's national culture.

However, an alleged symbolic dimension of destruction can also be used to justify reconstruction when it is neither provable nor even plausible. For example, Lithuanian historiography emphatically interprets the demolition of the dilapidated Palace of the Grand Dukes in Vilnius in 1799–1801 as a conscious annihilation of a symbol of Lithuanian statehood by the Russian occupiers.[14] Consequently, the controversial reconstruction of this building eradicated more than two centuries beforehand – a project often been derided by critics – has been justified by its supporters as compensation for historical injustice. Yet in this case, the insinuation of political, symbolic intent behind the demolition seems to be based on a twentieth-century mindset rather than on an understanding of the time around 1800. As a matter of fact, the demolition of the ruins of the palace was, for its time, neither scandalous nor even exceptional; it was fully in line with the common approach to architectural monuments which had lost their function. Until the late nineteenth century, it was common practice (not only in tsarist Russia) to demolish venerable ruins and even quite intact buildings which nowadays would be attributed great historical significance and high artistic value. In most cases, the reasons for demolition were in most cases of a functional, economic, or hygienic nature rather than iconoclastic. Nevertheless, the interpretation of the demolition of the Palace of the Grand Dukes as a foreign act of aggression against Lithuania's national identity has been an important tool for the popularization of the reconstruction project.

The next thesis reflects a pattern of the process of implementation of most reconstruction projects.

*4.    Most of the projects emerged from civil society. However, as soon as a project becomes promising and politically useful, it gains the support of state authorities. The rulers then use their power to have the project implemented and exploit it for their own purposes.*

An extreme example of this is the Cathedral of Christ the Saviour in Moscow. During the perestroika period, some intellectuals started drawing public attention to the fate of the cathedral, which had been almost completely forgotten after its demolition by dynamite in 1931. It was not long before a citizens' initiative began demanding its reconstruction. Its activists believed that rebuilding the cathedral could be an expression of collective regret for the Bolsheviks' frenzy of destruction and would enhance the moral and spiritual values of Russian society. However, after the collapse of the Soviet Union, Russia's President Boris Yeltsin turned the original bottom-up idea into an official state project. Moscow's mayor Yuri Luzhkov proceeded to use his authority to drive the reconstruction of the cathedral forward as one of the major building projects in the Russian capital and put pressure on organizations and the private sector to support it financially. A strategic alliance between the Russian Orthodox Church, the state and new Russian big business made sure that the cathedral was rebuilt with breathtaking speed and that priority was given to ensuring a sparkling appearance over attention to detail and historical accuracy. The enormous costs, the faked splendour, the involvement of shady businessmen, and not least the exploitation of the cathedral for nationalist propaganda and self-aggrandizement of politicians provoked harsh criticism from liberal intellectuals. 'The intelligentsia wanted faith. But it got the Cathedral of Christ the Saviour', one detractor sneered (Akinsha and Kozlov 165).

Reconstruction projects like this are a bone of contention for professional curators of monuments. Particularly in Germany, these circles tend to regard themselves as untouchable guardians of the principle of the irreplaceability of the original historical building, which since around 1900 has been one of the central doctrines of professional monument conservation. However, this self-conception is a fallacy, as the last thesis attempts to demonstrate.

5.    *Professional curators of monuments have often stressed their general criticism of the reconstruction idea. However, rarely have they firmly opposed reconstruction projects which were seen as particularly important for national identity or other emotional needs of society. In such cases they have often been inclined to make an exception to their principles.*

A good example of this flexible approach to the principles of conservation theory is Conrad Steinbrecht, who directed the restoration work at Marienburg during the decades around 1900. For his time, Steinbrecht was a very modern curator of monuments. He criticized the common restoration practice of the nineteenth century which produced idealized images of the past by purifying the shape of buildings and adding reconstructed elements. Instead, he advocated the maintenance of buildings in their authentic state of preservation, be it as a ruin.[15]

But not so in the case of Marienburg. Opting for total reconstruction, he did not hesitate to interfere in the castle's fabric in order to recreate its alleged medieval appearance. He argued that the centuries-long neglect of Marienburg, which had led to its dilapidation, reflected the neglect of German culture and German interests in this part of Europe. In his view, the huge castle complex was the essence of the German Teutonic Order, and even more so of the history of Germany's eastern territories. The symbolic, missionary significance of Marienburg as a stronghold of Germanness in the nationally contested eastern part of the German Empire was for him the main reason to justify its total reconstruction:

> This is a site where important memories of our Fatherland's history and countless strands of cultural work come together; this is a constant source of inspiration for patriotic sentiments … . In a nutshell, it is a creative edifice, and we should employ all available means to clearly restore it in a manner which is comprehensible to not only experts but also the masses, so that the Germans remain aware of their ancient rights to their contested homeland on the Vistula and their cultural superiority.[16] (Steinbrecht 406)

Although Steinbrecht normally propagated the central postulate of modern monument conservation theory of 'conservation, not restoration' ('konservieren, nicht restaurieren'), the national significance of Marienburg was for him a more important aspect which justified the exception to this rule. Similarly, Polish curators of monuments before and after the First World War generally supported the theory's anti-reconstructivist principles, but did not hesitate to disregard them if they deemed the reconstruction of a building necessary for national culture, such as Wawel Castle in Kraków and Staszic Palace in Warsaw.

This was especially true after the Second World War. Poland's conservator general Jan Zachwatowicz, justified the reconstruction of destroyed historical city centres with the following dramatic words: 'Our sense of responsibility to future generations demands that we rebuild the part of us that was destroyed, that we rebuild it completely, aware of the tragedy of the conservation forgery that we are committing'. To validate this, he borrowed Steinbrecht's argument and almost his wording, without of course explicitly referring to him:

> Architectural monuments are not only for gourmets; they are suggestive
> documents of history in the service of the masses. Even if they are deprived of
> their antiquarian values, they fulfil a didactic and emotional function. …
> The issue of monuments is a key issue of society – it is an issue of national culture.
> We cannot apply to this issue a one-sided abstract theory. We have to consider
> the needs of the present.[17] (Zachwatowicz 1946 52)

Given the immense emotional and political importance of rebuilding above all the systematically destroyed old town of Warsaw – but also because they had placed the alleged needs of the nation above their theoretical professional postulates long before the Second World War – Poland's conservationists quickly signed up to the reconstruction course despite isolated dissenting voices. And if any of them had any doubts, they could not discuss them publicly – at least not once the Stalinisation of the country intensified as of 1949.

Interestingly, even the curators of monuments in West Germany, who were famous for their strictly anti-reconstructivist attitudes, expressed some understanding for the Polish reconstruction campaigns and justified them given Poland's particular emotional situation after the Second World War. Torsten Gebhard, an influential official of West German monument conservation, wrote in 1958 on the reconstructed old town in Warsaw:

> From a historical, antiquarian viewpoint, the loss of building stock in Warsaw was
> so immense … that the old town has been irretrievably lost, making the recovery
> of the original impossible. However, Poland considered it necessary to rebuild,
> to reconstruct the whole fabric [of the city], in order to recreate the original
> experience. Theoretically, it would of course also have been possible to impose
> a completely newly built old town in contemporary architectural forms … . But
> it is part of the mindset of our time that we cannot abandon the values making
> up the diverse structures of our cities without a fight. … They represent our last
> emotional bonds to the world of our ancestors.

Gebhard not only defended the decision to reconstruct old Warsaw, he also praised its result: 'The spatial quality of the city has been definitely recovered … and it conveys emotional experiences. … Even a historian', he concludes, 'cannot say in this situation that the result of reconstruction does not mean anything to him … . He must acknowledge that the idea of the old town … has been revived'[18] (Gebhard 79–81).

Like other German commentators on the Polish projects in the early post-war years, Gebhard might have felt that it would be absolutely inappropriate for him to criticise his Polish colleagues for reconstructing cities which had been a victim of German aggression. But in West Germany as well, numerous buildings were reconstructed after the Second World War. Although the reconstruction projects provoked heated debates, it would not be true to say that all curators opposed them. In the face of the devastated cities, the old postulate 'conservation, not restoration' had become absurd.

Art historian and monument curator Hiltrud Kier described the situation retrospectively regarding the rebuilding of the Tower of Great St Martin's Church in Cologne: 'It was obvious – theoretically it was not allowed to rebuild the tower. But it was also obvious that practically it would be rebuilt'[19] (Hassler 327–8). The same applied to numerous other buildings of exceptional value, such as the other Romanesque churches in Cologne, Goethe House in Frankfurt, City Hall in Münster and St Michael's Church in Hildesheim, to mention only a few examples.

In a nutshell, West German curators of monuments adhered to their anti-reconstructivist principles, but only as long as it was not too painful. The same can be said of monument curators in present-day reunited Germany. Despite sharply criticizing the proliferation of reconstruction projects since 1990, hardly any of them are not willing to make an exception whenever a project is borne by a strong emotional need felt by a majority of the public.

This happened in Dresden when the Frauenkirche (Church of Our Lady) was reconstructed. The harsh criticism which accompanied the project during its beginnings in mid-1990s has faded. The rebuilt church has been enthusiastically welcomed by the majority of the city's population – and at the end of the day also accepted by monument curators as a legitimate exception to the rule.

In this case, the need for national self-assertion, which has been a driving force of reconstruction since the nineteenth century, did not play a significant role anymore. The same applies to other German reconstruction projects of the present, including the highly controversial Royal Palace in Berlin. In several post-socialist and, in particular, post-Soviet states, this need is still so vital that there is no argument which could stop nationally-motivated reconstruction projects, not even outdated principles of monument preservation.

## BIBLIOGRAPHY

Akinsha, Konstantin and Grigorij Kozlov. *The Holy Place. Architecture, Ideology, and History in Russia*. New Haven: Yale Universty Press, 2007. Print.

Bartetzky, Arnold. *Nation – Staat – Stadt. Architektur, Denkmalpflege und Visuelle Geschichtskultur vom 19. bis zum 21. Jahrhundert [Nation – State – City. Architecture, Heritage Preservation and Historical Culture in the 19th–21st centuries]*. Köln / Weimar / Wien. Böhlau, 2012. Print.

Boockmann, Hartmut. *Die Marienburg im 19. Jahrhundert [The Marienburg Castle During the 19th Century]*. Frankfurt a. M. / Berlin: Propyläen, 1992. Print.

Borger, Hugo, ed. *Der Kölner Dom im Jahrhundert seiner Vollendung [The Cathedral of Cologne in the Century of its Completion]*. Köln: Historische Museen der Stadt Köln, 1980. Print.

Bulkin, V. A. *Nowgorod: Architekturdenkmäler – Kunsthistorisches Museum: Illustrierter Reiseführer [Novgorod: Architectural Monuments– Art Museum. An Illustrated Travel Guide]*. Leningrad: Aurora-Kunstverlag, 1984. Print.

Dettloff, Paweł. *Odbudowa i restauracja zabytków architektury w Polsce w latach 1918–1939. Teoria i praktyka [The Reconstruction and Restoration of Architectural Monuments in Poland in the Years 1918–1939]*. Kraków: Tow. Autorów i Wydawców Prac Naukowych 'Universitas', 2006. Print.

Dettloff, Paweł, Marcin Fabiański and Andrzej Fischinger. *Zamek królewski na Wawelu. Sto lat odbudowy (1905–2005) [The Royal Castle on the Wawel Hill. Hundred Years of*

*Reconstruction (1905–2005)]*. Kraków: Zamek Królewski na Wawelu – Państwowe Zbiory Sztuki, 2005. Print.

Dolinskas, Vydas and Daiva Stepanovičienė, eds. *Lietuvos Didžiosios Kunigaikštystės Valdovų rūmai ir jų atkūrimas europinės patirties kontekste / The Palace of the Grand Dukes of Lithuania and its Restoration within the Context of the European Experience*. Vilnius: Kultura, 2009. Print.

Gebhard, Torsten. 'Zum Wiederaufbau von Warschau' [On the Reconstruction of Warsaw], *Deutsche Kunst und Denkmalpflege* 16.1 (1958): 79–81. Print.

Görres, Joseph. *Der Dom von Köln und das Münster von Straßburg [The Cathedrals of Cologne and Strassbourg]*. Ed. Bernd Wacker. Paderborn: Schöningh, 2006. Print.

Hassler, Uta, Winfried Nerdinger, eds. *Das Prinzip Rekonstruktion [The Principle of Reconstruction]*. Zürich: vdf, Hochschulverlag an der ETH, 2010. Print.

Hinterkeuser, Guido, ed. *Wege für das Berliner Schloss / Humboldt-Forum: Wiederaufbau und Rekonstruktion zerstörter Residenzschlösser in Deutschland und Europa (1945–2007) [Ways for the Berlin Palace / Humboldt Forum. The Rebuilding and Reconstruction of Destroyed Residential Palaces in Germany and Europe (1945–2007)]*. Regensburg: Schnell & Steiner, 2008. Print.

Jakimowicz, Teresa, ed. *Architektura i urbanistyka Poznania w XX wieku [Architecture and Urbanism in 20th Century Poznań]*. Poznań: Wydawn. Miejskie, 2005. Print.

de Keghel, Isabelle. 'Der Wiederaufbau der Moskauer Erlöserkathedrale. Überlegungen zur Konstruktion und Repräsentation nationaler Identität in Rußland' [The Rebuilding of the Cathedral of Christ the Saviour in Moscow. Reflections on the construction and Representation of National Identity in Russia]. *Inszenierung des Nationalen. Geschichte, Kultur und die Politik der Identitäten am Ende des 20. Jahrhunderts*. Ed. Beate Binder, Wolfgang Kaschuba and Peter Niedermüller. Köln: Böhlau, 2001. 211–32. Print.

Klein, Adolf. *Der Dom zu Köln. Die bewegte Geschichte seiner Vollendung [The Cathedral of Cologne. The Turbulent History of its Completion]*. Köln: Wienand, 1980. Print.

Kostílková, Marie. *Katedrála sv. Víta [The St. Vitus Cathedral], vol. 2: Dostavba [The Completion]*. Praha: Helios, 1994. Print.

Krauß, Jutta. *Die Wiederherstellung der Wartburg im 19. Jahrhundert [The Rebuilding of the Wartburg Castle in the 19th Century]*. Kassel: Wartburg-Stiftung, 1990. Print.

Kubiček, Alois. *Betlémska kaple [The Bethlehem Chapel]*. Praha: Státní nakladelství krásné literatury, hudby a umění, 1960. Print.

Nerdinger, Winfried, ed.. *Geschichte der Rekonstruktion – Konstruktion der Geschichte [History of Reconstruction – Construction of History*. München: Prestel, 2010. Print.

Steinbrecht, Conrad. 'Die Wiederherstellung des Marienburger Schlosses' [The Rebuilding of the Marienburg Castle]. *Centralblatt der Bauverwaltung* 16 (1896): 36–7, 397–9, 405–6, 411–13. Print.

Zachwatowicz, Jan. 'Program i Zasady Konserwacji Zabytków' [The Programme and the Principles of Monument Preservation]. *Biuletyn Historii Sztuki i Kultury* 8. 1–2 (1946): 48–52. Print.

Zachwatowicz, Jan. *Ochrona Zabytków w Polsce [Monument Preservation in Poland]*. Warszawa: Polonia, 1965. Print.

Zachwatowicz, Jan: Program i zasady konserwacji zabytków [The Programme and the Principles of Monument Preservation]. *Biuletyn Historii Sztuki i Kultury* 8, no. 1–2, 1946, 48–52. Print.

## NOTES

1   This text is largely based on a recent article in German: Bartetzky, Arnold. 'Das Denkmal und Seine politische Bedeutung. Rekonstruktion und Nationenbildung'. *Werte. Begründungen der Denkmalpflege in Geschichte und Gegenwart.* Ed. Hans-Rudolf Meier, Ingrid Scheurmann and Wolfgang Sonne. Berlin: Jovis, 2013, 232–45. For further information on the reconstruction projects discussed, see also the chapters '"Wiedergutmachung für historische Verluste". Der Wiederaufbau von Baudenkmälern im östlichen Europa als Akt nationaler Selbstbehauptung' and '"Seid von Zeit zu Zeit auch tolerant!" Historische Positionen der Denkmalpflege zur politisch motivierten Rekonstruktion zerstörter Baudenkmäler', in Bartetzky 2012, 17–52. I am indebted to Chris Abbey and Louise Bromby for the linguistic revision of this text.

2   'ein Siegesdenkmal für das alte Ordensland, das sich so gern rühmte, die anderen Deutschen zum heiligen Kampfe erweckt zu haben'.

3   'Sie werden der Erste sein, [ … ] der dem Volke den Schrein öffnet, in dem es die Dokumente seiner Größe findet, [ … ] und von der Erbauung der Wartburg wird der Deutsche einst die schöne Epoche seiner Selbsterkennung datieren'.

4   'die Kräfte Teutschlands zur Vollendung [zu] verbinden'; 'Schande und Erniedrigung'; 'eigenem Hader und fremdem Uebermuthe' [preisgegeben]; 'ein Symbol des neuen Reiches, das wir bauen wollen'.

5   'V době, kdy se Praha stala nevýznamným provinčním městem a tuhý centralismus a germanizace potlačily její kulturní i společenský život, nedokončená katedrála připomínala slavnou minulost českého království … a vzbuzovala představy o své dostavbe'.

6   'monumentu wielkości i kultury narodu'.

7   ' … domaga się jej sumienie narodowe'.

8   'Bethlehem, aus dem der tschechische Messias aufgehen wird'.

9   'zur ersten Volkstribüne in der Welt [gemacht]'; 'der Ursprung der Hussitenbewegung, der größten revolutionären Etappe der tschechischen Geschichte'.

10   'Rückgewinn eines für das nationale Selbstbewusstsein und die geschichtliche Erinnerung gleichermaßen wichtigen Symbols'; 'Ausdruck der staatlichen Souveränität Litauens'.

11   '[ … ] że cały szereg obiektów został zniszczony nie przypadkowo, w bezpośrednich działaniach wojennych, lecz jako świadomy akt likwidacji dorobku kultury polskiej, dokonany przez hitlerowców'.

12   'Nie mogąc zgodzić się na wydarcie nam pomników kultury będziemy je rekonstruowali, będziemy je odubowywali od fundamentów, aby przekazać pokoleniom, jeżeli nie autentyczną, to przynajmniej dokładną formę tych pomników, żywą w naszej pamięci [ … ]'.

13   'Während des Großen Vaterländischen Krieges 1941–1945 hatten die faschistischen Barbaren die Absicht, Nowgorod dem Erdboden gleichzumachen. [ … ] Die Außerordentliche Staatliche Kommission zur Untersuchung der Greueltaten der faschistischen Okkupanten stellte fest, daß der von der Stadt erlittene Schaden über 1 Mrd. Rubel betrug'.

14   Hinterkeuser 197. See also Dolinskas and Stepanovičienė 122–4.

15   Steinbrecht 405–6.

16    'In dieser Stätte vereinigen sich viel wichtige Erinnerungen vaterländischer Geschichte und zahllose Fäden cultureller Arbeit; von hier gehen stete Anregungen für patriotischen Sinn [ … ] wieder aus. Es ist mit einem Wort ein Schöpfungsbau, und den müssen wir uns mit allen Mitteln handgreiflich wiederherstellen: nicht bloß verständlich für den Kenner, sondern anschaulich für das Volk, damit das Deutschthum auf dem strittigen Boden an der Weichsel sich seines älteren Heimathsrechtes und seiner höheren Culturaufgaben bewußt bleibt'.

17    'Poczucie odpowiedzialności wobec przyszłych pokoleń domaga się odbudowy tego, co nam zniszczono, odbudowy pełnej, świadomej tragizmu popełnianego fałszu konserwatorskiego. Zabytki bowiem nie są potrzebne wyłącznie dla smakoszów, ale są to sugestywne dokumenty historii w służbie mas. Wyzbyte wartości starożytniczych, będą nadal pełniły służbę dydaktyczną i emocjonalno-architektoniczną. [ … ] Sprawa zabytków jest podstawowym zagadnieniem społecznym – zagadnieniem kultury narodu. Nie możemy wobec nich stosować jednostronnie abstrakcyjnej teorii, musimy umzględnić potrzeby dnia dzisiejszego'.

18    'Vom historisch-antiquarischen Standpunkt ist der Substanzverlust in Warschau so groß [ … ], daß die Altstadt unwiederbringlich für eine Wiederherstellung des Originalen verloren ist. Polen hat es aber für erforderlich gehalten, den Gesamtkörper gerade im Hinblick auf seinen Erlebniswert – [ … ] zu rekonstruieren. [ … ] Es gehört [ … ] zur geistigen Gesamtstruktur unserer Zeit, daß wir nicht ohne weiteres auf derartige Werte verzichten können, wie sie diese ungemein differenzierten Gesamtgebilde unserer berühmten Altstädte darstellen. [ … ] Es geht hier [ … ] um letzte seelische Bindungen an die Welt der Vorfahren. [ … ] Selbst der Historiker kann angesichts dieser Verhältnisse nicht erklären, das, was der Wiederaufbau zustande gebracht habe, sei für ihn unergiebig, da nur ein verschwindender Prozentsatz an originaler Substanz mit hinübergerettet wurde. Er muß zugeben, daß die Idee Altstadt [ … ] wiedererstanden ist'.

19    'Es war klar, theoretisch durfte man den Turm nicht wiederaufbauen, und es war klar, daß er praktisch wiederaufgebaut wurde'.

# 2

# Barcelona's Gothic Quarter: Architecture, Ideology and Politics

*Josep-Maria Garcia-Fuentes*

Nowadays, when presentation seems to be everything, Barcelona's Gothic Quarter or 'Barri Gòtic' could be taken for an authentic, carefully preserved, medieval district. Despite this, however, it was devised and built between the 1920s and 50s and not in the thirteenth, fourteenth or fifteenth centuries as one would be led to believe by its name – which was conceived along with the quarter itself. As a result of this construction process, Barcelona's Gothic Quarter, as it is known to this day and which has barely changed since its consolidation in the 1960s, has imprecise boundaries. All told, it is rather difficult to define. It is not an authentic quarter and neither is it a restoration, a 1:1 reconstruction or a fake, although its construction was indeed inspired by the idealization of certain Gothic buildings. The Barri Gòtic stands halfway between all these things. It is at once both completely authentic and completely artificial. It is an imaginary and idealized space which did not previously exist but which was created using real Gothic stonework and buildings reconfigured around seven real Gothic structures – including Barcelona's cathedral, the Royal Chapel of Santa Àgueda and the Tinell Hall.

Its construction did not entail the preservation of an inherited architecture or the restoration of a real Gothic Quarter which was on record or which it was possible to rebuild. It was and still is the outcome of a complex and effective architectural, cultural, political, social and touristic struggle and negotiation in which the quarter was built according to an idealization of its own. That is to say, old Gothic stones, new construction materials and existing places were re-contextualized in the Gothic Quarter with the aim of creating a new idealized narration of the city and its history, which has undergone constant change as a result of this complex process of struggle and negotiation.

## ORIGINS

The invention of Barcelona's Gothic Quarter, as Joan Ganau[1] pointed out, was brought about by the construction of the cathedral's new façade and by the intense struggle to renew and expand the city. Both these processes took place in the last half of the nineteenth century and both of them were closely tied to the search for an idealized Catalan-Spanish national architecture and to the birth of a bourgeois Catalan nationalism that shaped the city and the identity of Barcelona as we know it today.

On the one hand, the proposal for a new façade was put forward in 1848 by one of the first Catalan Romantics, Pau Piferrer, who also introduced modern medievalism in Catalonia and promoted the switch to a medieval taste. Noting that Barcelona's cathedral remained 'unfinished', Piferrer proposed its completion in the same style in which it had been conceived, that is, the Gothic. The discovery in the cathedral's archives of what was purported to be the original drawing of its Gothic frontispiece –drawn by the master builder known as Mestre Carlí – proved to be a decisive factor in the social consensus on Piferrer's idea. The works were funded by the local banker Manuel Girona and they began in 1860 according to a project drafted by the architect Josep Oriol Mestres on the basis of Mestre Carlí's original drawing.

Due to the conflictive political and social conditions in Spain and particularly in Barcelona at that time, however, the works suddenly halted soon after that date and were not resumed until the 1880s, when the project finally started up again, now in a period in which society had come to acquire Romantic tastes. Not surprisingly, disagreement arose regarding the type of Gothic to be used. Various possibilities were considered: some wanted the design to be strictly based on the purported 'original drawing' discovered in the cathedral archives, as was the case of the project by Oriol Mestres – the option supported by Manuel Girona. Other projects, however, wanted to completely reinvent the façade according to the contemporary Gothic idealization upheld by Viollet-le-Duc – a constant reference for all the Catalan architects at that time – in keeping with the contemporary projects in Gothic architecture that were being carried out in France and Germany. These projects were grounded in the idealization of Gothic architecture and in the final form envisaged for the building rather than in the 'original' architecture, and this was something that brought them more closely into accord with the political, social and cultural aspirations of the new local bourgeoisie and their goal of renewing the city and building a new modern, monumental and symbolic universe for it. This group of projects included, for example, the one designed by the architect Jeroni Martorell and drafted by the young Antoni Gaudí and Lluís Domènech i Montaner, a proposal which featured the construction of an imposing dome in Viollet-le-Duc's Gothic style. It should be noted, however, that the original drawing by Mestre Carlí only defined the frontispiece, that is to say, the doorway of the cathedral. The rest of the façade remained to be invented by the architect. Consequently, both types of project interpreted the past creatively according to present needs.

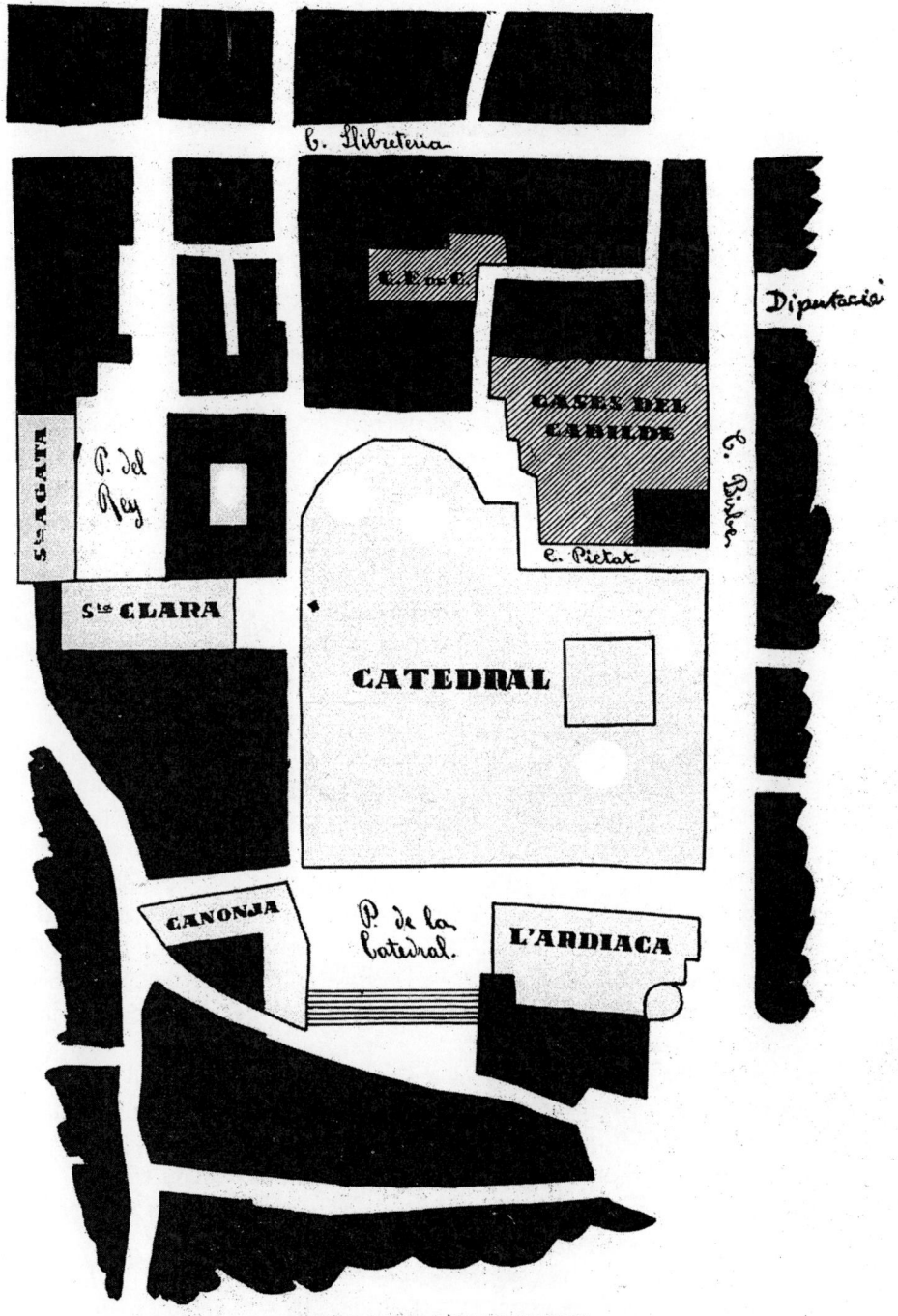

PLA DEL «BARRI GÒTIC» INTANGIBLE
(L'assenyalat en negre *no* és gòtic.)

2.1   1927 map of non-Gothic buildings (black) in the Gothic Quarter

Both defined the new façade through a creative reinterpretation of the past, regardless of whether the project was a Gothic elaboration seeking legitimacy on the basis of the 'original' drawing discovered in the cathedral, or a 'modern' concept based on an ambitious Gothic invention. Ultimately, to a greater or lesser extent, both were invented solutions. In other words, the two approaches were not opposites but rather they sought to define and idealize Catalan Gothic architecture as an expression of the nation that was being forged. In this context, reinventing the cathedral in a new idealized Gothic style was understood by the local elites to entail the recovery of the medieval Barcelona's magnificence and of the splendorous times when the city had held sway over trade in the Mediterranean. Beyond the struggle and differences between the various positions,[2] this reveals the common representative ambitions of the local cultural and social elites to shape the city as a cultural and economic capital.

Yet the problem was not just a matter of representation or of monumentalization of the city, but one of congestion as well. The harsh military oppression to which the city had been subjected since the time of the War of the Spanish Succession (1701–14) continued to forbid the expansion of Barcelona outside its city walls during the early years of the local industrial revolution in the first half of the nineteenth century. This situation led progressively to an excessive densification of the city, which had become totally congested by the 1850s, making it urgently necessary to carry out an expansion in order to avoid health and infrastructure problems (the city's sewer and water supply systems were completely lacking) as well as for functional reasons.

In this context, of course, the representative ambitions of the local elites to shape Barcelona as a cultural and economic capital also set the course of the discussion about the expansion and renewal of the city. So, while the local authorities and the local cultural elites considered this to be a great opportunity to re-conceptualize the whole city according to their own particular ambitions, the Spanish authorities – who actually held the power to make the decision – saw the situation in purely functional terms (Guàrdia and Garcia-Fuentes 47–95). Despite that, the fact is that the city walls came to be demolished in 1854, and in 1859 the City Council, taking up the problem as a matter of its own responsibility, decided to hold a competition for ideas to face the issue.

It is interesting to take a look at the proposals submitted in the competition because they clearly show how the cathedral played a key role in the various monumentalizations of the city that they envisaged. All the projects included new boulevards and broad avenues with the aim of connecting the new expansion to the existing city. These thoroughfares were plotted to run from new squares located in the planned new urban developments to another square – of varying size depending on the proposal – in front of the cathedral's new façade. This is what may be seen, for example, in the project by Antoni Rovira i Trias, the winner of the competition.

Indeed, this is also what can be seen in the project by Ildefons Cerdà that finally prevailed, although Cerdà's conception was less monumental.[3] His project considered both the new expansion of the city – Barcelona's celebrated Eixample – and the renewal of the existing city. In this respect, he planned for the building of three new straight avenues that would transect the Old City in order to sanitize and regenerate it as well as to connect the existing areas to the new expansion district. These new avenues were called Via A, Via B and Via C. The first two were aligned from the Collserola mountain range to the seafront, coinciding with the expansion district's famous grid of streets. The third avenue, Via C, ran almost parallel to the cathedral's façade. This highlights Cerdà's awareness of the importance of the cathedral's new Gothic façade for the local inhabitants, since even his functionally-oriented project took the cathedral into account on planning the renewal of the existing city.

Neither of these processes – the city's renewal and expansion and the cathedral's 'completion' – were unique to Barcelona but rather a common contemporary trait in many European cities. In all cases the proposals involved the destruction of extensive spaces in the existing cities that were significant in terms of memory and ancestry, as the architect Lluís Domènech i Montaner stated in 1879 in an article about the renewal of Barcelona published in the Catalan newspaper La Renaixensa. Despite the trauma caused by this destruction, Domènech was totally convinced of the need to take up the challenge of improving the city by reshaping it creatively, and he proposed that a new radial structure should be established for the future city with its centre at the old Roman Acropolis, the cradle of the first Barcelona in history, next to the cathedral's apses. To this end he proposed the demolition of all the houses in the area between Plaça Sant Jaume and the apses, in order to define a new large square in which the entire history of the city would be represented and condensed: the three surviving columns of the Roman forum, the Bishop's palace, the palace of the Generalitat and the Gothic cathedral. The memory of Barcelona would be represented in this new space and it would be possible to contemplate the city's entire history there. In the words of Domènech i Montaner, this would be a unique square '[ … ] straightforward, complete, and more replete with majesty, memories and beauty than is to be seen in any other city worldwide' (125–48).

In this way, Domènech's proposal reinvented the cradle of Barcelona by interpreting its entire past and memories – and not just its Gothic architecture – in a creative way with the aim of modernizing the city according to its present needs. It was the proposal for a new Barcelona built on the selective destruction of the old urban landscape, seeking to evoke the ancient splendour of the Roman colony of Barcino on the same site where it once stood.

The proposal soon became widely known and accepted – revealing in this way the late nineteenth-century contradiction between a predilection for medieval architecture, on the one hand, and a taste for classicist urbanism and a rigid Baroque scenery closer to the Beaux-Arts tradition on the other (Ganau 1997). Indeed, a significant reflection of the wide acceptance of Domènech's idea is found in its description in the famous poem Oda a Barcelona (Ode to Barcelona) by the great Catalan poet of that period, Jacint Verdaguer, who wrote in 1883:

> Saint George of the ruling Court seeks to see Saint Claire
> the Counts' ancient palace yearns for that of the noble Council …
> Oh! Tear down the curtain of houses separating

*King James' statue from his royal Tinell Hall.*
*In the middle of that square, which will have no peer,*
*the traveller, catching sight of Hercules' three columns*
*will think he sees Three Graces, in crowning glory,*
*arms linked, dancing in your garden.[4]*

This poem by Verdaguer, which includes Domènech's vision of a future reinvented city, became extremely popular. The City Council published 100,000 copies in 1883 alone and the proposal was included by Àngel Josep Baixeras in a new version of his project for the renewal of the Old City in that same year. Even in 1911 – 28 years later and in a completely new political context – the proposal was still alive and was finally drawn by the architect Francisco de Paula Nebot in a view entitled *Project for a Square at the Acropolis of Barcelona* – this time, however, with a very different aim as shall be seen further on.

Although none of these projects came to be a reality, they do provide a valuable insight into the cultural and social circumstances surrounding the invention of the Gothic Quarter. They clearly show that the cultural and symbolic reinvention of the city (and especially of its monumental centre) in accordance with its citizens' contemporary needs and ambitions, was one of the major concerns and wishes of Barcelona's local elites for many years.

## THE IDEA

The first time the idea of building a Gothic Quarter came up was in 1908 when demolition work to make way for the Via A (the first of the avenues proposed by Cerdà to renew and improve the city centre) reached its peak.[5] As expected, the works brought to light a large number of Gothic stones that were considered of interest and even entire buildings. Indeed, the amount of stonework that was found surpassed even the most optimistic predictions: Gothic windows, arcades, stairs and in some cases complete structures that had been hidden by centuries of constant rebuilding were revealed. The City Council, which had reserved the right to retrieve any significant or valuable stonework in a specific contract with Banco Hispano Colonial (the bank that financed the works), was overwhelmed, none of the city's museums being able to house such quantities of new material (Ganau 2006 11–23). The municipal warehouses were soon filled up to capacity and unable to take in any more material, so one of the ways it was decided to preserve the findings was to relocate ancient buildings to the city's new expansion district or to other areas, as was the case of the Coppersmith's Guild House (Gremi de Calderers), relocated temporally to Plaça Lesseps in the northern part of the city. Actually, this idea was not completely new. Although the circumstances differed, in late nineteenth century a few buildings from the Old City had been relocated to new locations. This happened to the Concepció cloister and church, for example, which had belonged to the Jonqueres monastery, and to Sant Ramon de Penyafort church. Thanks to these relocations it was possible to preserve entire buildings and to prevent the destruction of their stonework while at the same time allowing the creation of a new symbolic and monumental network in the new areas of city growth, which were eager for symbolism.[6]

However, the brusqueness and aggressiveness of these relocations and especially the great destruction caused by the demolition works making way for the Via A caused fierce struggles between their supporters and detractors. For those who were against the works, it was not enough to preserve the most valuable buildings in a new location and even less acceptable to merely preserve isolated stonework features. They argued that it was also necessary to preserve and enhance the streets and their atmosphere, that is to say, to preserve the urban fabric. This new appreciation was the result of the acceptance of the ideas of Camillo Sitte and Charles Buls, ideas that spread through Catalonia and Spain at large in the opening years of the twentieth century. This was a new awareness that also influenced the City Council's initiatives before the demolition works began: an artistic competition to represent the Old City was organized, in which artists and photographers were asked to leave a testimony of the Barcelona that was about to vanish, and the historian Francesc Carreras i Candi was commissioned to write the history of the old district just before it came to be completely transformed and redeveloped (Torrella 128–47).

Nevertheless, the demolition works did not mark an end to the process. Once the peak of destruction had been reached, it was necessary to think of how to repair the urban wound. Doubts clearly existed in the face of so much destruction: how should the new buildings in the Via A be built and in what architectural style? Moreover, what should be done with all the valuable material salvaged from the demolitions?

In an interview with the newspaper *Diario de Barcelona* in November 1908, the architect Antoni Gaudí stated that the writer Joan Maragall was working on a manifesto in favour of the Old City that would be supported by all the cultural associations of Barcelona. Gaudí went on to advocate the preservation of the stones recovered from the demolition process and their use to improve the city's appearance, and even sketched some possible embellishments. A month later, in December 1908, the Artists Association of Barcelona and the Architects Association of Catalonia publicly presented two documents in which they proposed to reconfigure and rebuild, close to the cathedral, the materials retrieved from the demolition works with the aim of 'forming a complex that sums up the art of ancient Barcelona [ … ]' (Nicolau and Venteo 116). The idea of recreating a medieval quarter around the medieval cathedral and its new Gothic façade, which was located very close to the modern Via A or Via Laietana, where the materials were coming from, was taking shape.

Finally, in 1911 the journal *La Cataluña* published a special issue on Barcelona's renewal with the participation of architects like Josep Puig i Cadafalch, Jeroni Martorell and Joaquim Manich, who contributed some views and a small urban plan. The paper also included an article by the writer Ramon Rucabado entitled 'A Gothic Quarter in Barcelona'. In it Rucabado clearly proposed the idea of building a Gothic Quarter – 'instead of a modern one (Rucabado 305)' – in the area around the cathedral, which was also adjacent to the new thoroughfare (Via A) created by the large-scale demolition works that had brought to light the Gothic stones.

The introduction to the special issue was signed by the architect Jeroni Martorell and it stated very clearly that it was necessary to reform the Via Laietana project before starting to construct its new buildings with the aim of defining an appropriate relationship between new buildings and the old architectures. The principles and ideas announced by Sitte and Buls, among others, provided general guidance on how to face the challenge, by seeking to achieve a unity of atmosphere and style – which should be Gothic, of course! Furthermore, Rucabado proposed that a unity of commerce and retailing should be achieved, even suggesting that the quarter should be devoted specifically to the second-hand book business. Once again, however, the discussion involved the question of how to plan the relocations and reconstructions. An idealized Gothic model that would resolve this matter had to be defined.

This theming or zoning proposed by Rucabado, far from being an isolated idea, was common in the planning of contemporary cities and in the contemporary discussions on Barcelona's urban plans. Accordingly, at the same time as the Via Laietana (located right next to the cathedral and the future Gothic Quarter, as previously mentioned) was planned as the financial and business district of the city, on the other side of Les Rambles the myths of District V and of the city's 'Chinatown' (Barri Xino) were being forged. Indeed, that was also the period when two young architects, Antoni Puig Gairalt and Lluís Bonet Garí, proposed the construction of a 'Baroque Quarter' around the Boqueria Market with the stonework from the future demolitions for the Via B.[7] Consequently, the proposal to build a Gothic Quarter in the area around the cathedral was based on the fact that there were a small number of 'Gothic' buildings in the area, the main one being, of course, the cathedral with its newly built 'Gothic' façade. One could be tempted to think that this proposal stood in contradiction to the contemporary architectural proposals of La Lliga (the leading political party in Catalonia at that time), set within the framework of the Noucentista movement which, in simplified terms, may be said to have advocated a neo-classical aesthetic in new buildings and urban spaces. This was not the case, however, because both strategies sought to define a genuine Catalan identity.

Nevertheless, as previously mentioned, it was also in this year that the architect Francisco de Paula Nebot illustrated the alternative regeneration project that included a large square 'at the Acropolis of Barcelona' (the area behind the cathedral), planned on an extension of the huge demolition project for the Via A. This contradiction should come as no surprise since it merely highlights the political division and equal balance in the City Council between the Catalanistas or advocates of Catalan independence – who wanted to build a Gothic Quarter – and the Republicans – who were supporting the plans by Nebot.

However, neither of these groups managed to impose their project until 1924 when, following the advent of the Primo de Rivera dictatorship in 1923, the new City Council tried to impose Nebot's project. The popular opposition (mainly from the Catalanistas and from the people who were opposed to the dictatorship) was so great that the idea had to be withdrawn. One of the most aggressive responses was written by the architect Puig i Cadafalch and published in *La Veu de Catalunya*. In it he stated that the project represented the 'destruction of the Acropolis of Barcelona'. Puig described the proposal as a 'quick sketch, something drawn but not meditated' and he even added a reference to Verdaguer's poem – which was quoted in the plans of Nebot's drawings – saying that it was a 'frightening project by a poet who has no feeling for architecture due to the inevitable opposition between visible forms and literature' (Puig i Cadafalch 1924 7).

After that, in 1927, following a few small 'restoration' interventions on Gothic buildings in the area around the cathedral and the construction of some new ones, the governor of Barcelona requested the official provincial architect Joan Rubió i Bellver to draw some 'visions' of the ancient Acropolis area, including all the existing Gothic monuments restored, 'finished' and free of additions, with new surroundings planned to highlight them. These 'visions' consisted of three panels that were displayed in the cathedral cloister: two of them were views of the whole complex as seen from the newly built Via Laietana, while the third one showed a sequence of interior views and a sketchy map of the area with a possible master plan of the proposal. Accordingly, the panels and the proposal by Rubió i Bellver, commissioned by the governor of Barcelona, recovered the idea of building a Gothic Quarter around the cathedral. The results were truly spectacular since it completely redid the district, mixing old and new Gothic features and buildings with even the ancient Roman columns and some new colourful roofs inspired by Gaudí's architecture.[8] Despite its virtues, however, the proposal generated an even greater, more intense and more interesting debate. Indeed, this debate was so passionate that the three panels of the 'visions' were assailed with acids and other caustic fluids while on display at the cathedral cloister. Many contemporary Catalan public personalities, like Carles Riba, Folch i Torres, Rovira i Virgili, Valls i Taberner, Ràfols, Bonaventura Bassegoda and others wrote harsh critiques that were even collected in a special issue of the journal *El Imán*, together with a very brief defence of Rubió.[9] It is interesting to note how almost all the critiques severely condemned the excessive intervention proposed by Rubió and principally his proposal to build new spires on the cathedral and new colourful tiled roofs. While some of the critiques were focused perhaps on the project, almost all of them were overly aggressive, showing that their authors did not understand the proposal correctly. In short, they may be considered criticisms of the architect himself – at that time Rubió was an old professional representing a phantasmagoric world of the past – and above all attacks on the dictatorial 'official' government (Solà-Morales). Indeed, the commissioning of these 'visions' was nothing other than an attempt by the dictatorship to appropriate the invention of the Gothic Quarter.

Once the largest public tensions had diminished, Rubió i Bellver wrote a pamphlet entitled *Taber Mons Barcenonensis*.[10] He chose a photograph of Barcelona's cathedral before the construction of its new Gothic façade to illustrate the cover. In the text, he systematically replied to some of the criticisms and argued that there is 'no need to respect [the Gothic Quarter] to any greater or lesser degree because it simply does not exist' (44). He also sought to demonstrate the nonexistence of any purported Catalan Gothic style while defending his Gothic proposal and its

2.2   Postcard
of the imaginary
project for a Gothic
Quarter by Rubió
i Bellver, 1927

appropriateness by considering it in an international context. In this way he tried
to defend himself from the aggressive criticisms of the type of Gothic models
that had been used to define the views, and accused him to ground his 'mistaken
invention' in an idealized Gothic style based on northern models and not in the
Catalan Gothic style.

## THE CONSTRUCTION

The proposal was halted by this intense debate until 1927 when, as a consequence
of the concern for the Old City, the Service for the Conservation and Restoration of
Historic Monuments was created, under the direction of the architect Adolf Florensa.
The decision to organize this body arose from a debate on the preparations for the
1929 Barcelona International Exposition and the possibility of building a replica of
the medieval Barcelona on the Exposition's grounds.

   This idea was inspired by a series of popular medieval replicas of cities and
architectures that had been built in several previous world's fairs. In the case of
Barcelona and its medieval city, however, this proposal was finally ruled out,
arguing that it made no sense to build replicas of existing buildings, such as
Barcelona's cathedral. It was consequently proposed to create the Service for the
Conservation and Restoration of Historic Monuments with the aim to improve
the existing spaces in the Old City and to enhance the surroundings of the few
medieval buildings there. At the same time, however, it was built the Spanish
Village or 'Poble Espanyol' on the exhibition grounds on Montjuïc hill, comprising
a whole set of replicas of popular architectures selected from different locations
in the Iberian Peninsula. The aim of this strategy would appear to be quite clear:
it was an attempt to present all these different architectures as a unified whole
in spite of their differences. That is to say, it sought to represent the 'unity' of

'Spanish architecture'. Of course, this also reshaped the symbolism associated with the related enhancement of Barcelona's medieval district around the cathedral.

Indeed, as Agustín Cócola Gant[11] states (2014 22–6), this strategy was not a mere idea but rather a series of complex parallel initiatives that succeeded in packaging the city's medieval assets – and especially the quarter around the cathedral – together with the brand new Spanish Village and other features as the main touristic attractions in the branding of Barcelona and the promotion of the city's image.

The Service for the Conservation and Restoration of Historic Monuments was thus created and Adolf Florensa – with the assistance of the historian Agustí Duran Sanpere – worked silently but effectively on the construction of Barcelona's Gothic Quarter almost up to his death in 1968. During these years Florensa improved streets, squares, corners and buildings, and although he coordinated and designed local interventions without following a master plan, all these interventions followed a clear strategy that finally succeeded in creating the quarter as we know it today. Among the key aspects guiding the project, a vague aim to preserve and enhance the 'atmosphere' was the most important. Florensa himself included it among the three key ideas of his work in the quarter, namely: to restore the existing buildings, to relocate valuable old structures into the quarter, and to enhance their overall worth by 'harmonizing the bland' (Florensa 1958 18). In accordance with this goal, a few entire buildings were carefully relocated into the quarter, making some slight changes in the composition of their façades. This was because, even without stating it clearly, the reconstruction works were actually driven by an idealized domestic or civil Catalan Gothic architecture that was inspired by what the architect Josep Puig i Cadafalch called the 'Catalan house' (1913 1041).

According to this idealization, every 'Catalan house' was a typical and traditional representative medieval Catalan construction.[12] It was characterized by the composition of its façade – mainly formed by flat roofs with galleries, towers often on the corners, and especially the stonework of the windows in their different styles together with a rough organization of the spaces around representative central entrance courtyards with a great arched entrance door. This idealized typical model of a 'Catalan house' was progressively defined by the Catalan architects searching for a Catalan national architecture, like Puig i Cadafalch or Lluís Domènech i Montaner, among others, as a reinterpretation of medieval models. These architects placed great emphasis on an idealized civil Catalan Gothic architecture, grounded in the definition and description of the Aragonese Style in Viollet-le-Duc's *Dictionnaire*, and in Camille Enlart's description of the Palace for the Provincial Council of Perpignan (related to Viollet-le-Duc's description as well), which showed how the uniqueness of the medieval Catalan Gothic architecture resided in the civil and domestic structures, that is, the traditional rural house and the palace, rather than in religious buildings. These ideas were first stated by George Edmund Street in his book *Some Account of Gothic Architecture in Spain*, published in 1865. Street identified a continuum behind the Roman and the Gothic house, highlighting the fact that the windows changed in medieval times with the use of double and triple traceries.

Grounded in this imaginary but without making any reference to it, the architect Florensa carried out a systematic building activity in the quarter around Barcelona's cathedral as well as in Montcada Street, near the Gothic cathedral of Santa Maria del Mar. One of his first interventions was the rebuilding of the Casa Padellàs, which now shelters Barcelona's City History Museum at Plaça del Rei. This Gothic building, a fifteenth-century palace, was originally located in the area that was demolished to make way for the Via Laietana. The building was dismantled during the demolition works and then kept in storage until 1930, when it was rebuilt in its present strategic location.

As Florensa explained, the construction of the Gothic Quarter was a long process. Like a passionate collector, Florensa worked for decades to improve many buildings with the stones and fragments stored in the municipal warehouses. However, he also managed to recover many of the Gothic buildings scattered about the city over the course of the nineteenth century during the renewal of its central area. He planned their transfer and carefully designed their new position within a growing Gothic Quarter that was constantly being reshaped by new interventions and additions.

This was a patient and tireless effort like that of the collectors who formed the Renaissance cabinets of curiosities, in which the constant re-contextualization of fragments and the emotions evoked by them are more important than their authenticity or than the truthfulness of their reconstruction. This was also similar to the patient always-changing works of the architect Sir John Soane in his house-museum at Lincoln's Inn Fields in London during the late eighteenth and early nineteenth centuries. In all these instances, the constant re-conceptualization of the various items or architectural fragments and the emotions which they evoke are more important than their authenticity or than the historical truth of their reconstruction.

This idea is clearly expressed in the collage that Català Pic made in 1935 for the journal of the Sociedad de Atracción de Forasteros (Society for Attraction of Foreigners) (Figure 2.3). In it one can see the Gothic Quarter as a collage in which the buildings are placed next to one another. Corners, partial views, buildings, stairs, stones and houses are recomposed with the aim of creating something completely new, something that had never existed before: the Gothic Quarter. Buildings from the fifteenth century merge with reconstructions from the twentieth century and jutting out above them all are the bell towers and spires of the cathedral. Taken as a whole, they form the Quarter's skyline. The composition of the collage is framed by two objects placed in the foreground: a Corinthian column that appears to recall the Roman origin of the place, and a Gothic cantilever depicting a lion's head that, as if it were a chimera,[13] seems to decry the modern construction of the whole quarter. In short, it portrays the Gothic Quarter as an urban collage of stonework features and even entire buildings, new and old – as a new imaginary space that had never before existed. What does this collage represent if not the essence of the quarter that Florensa was building? Once again, the past was interpreted and idealized in a creative way. In the 1930s, however, not everyone agreed with this strategy. The group of young modern architects called GATCPAC

2.3    Collage by Català Pic for *Sociedad de Atracción de Forasteros* magazine, 1935

(Group of Catalan Artists and Technicians for the Progress of Contemporary Architecture) were strongly opposed to this kind of intervention in the historic city, which they qualified as 'false historicism'. At the same time they revisited the 'real' historic Gothic structures, seeking to learn some of the principles that should lead to a new modern and locally-rooted architecture. These architects worked on a master plan for District V and Barcelona's 'Chinatown', exploring these ideas in depth, and drafted a now famous and quite ambitious proposal called the Macià Plan for the overall renewal of Barcelona's historic city and its expansion area. This plan, which was commissioned by the Catalan government, was developed in collaboration with Le Corbusier. On considering these proposals, one finds that GATCPAC postulated extensive selective demolitions in the Old City with the aim of sanitizing this part of Barcelona and bringing it up to the same modern standards than were planned for the city's new expansions. Even so, they also planned to preserve the most consolidated streets and existing buildings in the Old City and around the cathedral together with their urban and architectural atmospheres. In any case they did not envisage rebuilding, enhancing or completing the existing buildings and areas as opposed to what had been planned by Florensa and the Service for the Conservation and Restoration of Historic Monuments – indeed, they were explicitly aggressive in their opposition to such interventions.

The Macià Plan never became a reality due to the political struggle in the 1930s and GATCPAC's criticism did not affect the progress of Florensa's work. It is interesting to note, however, that after the Spanish Civil War (1936–9), the works to build the Gothic Quarter resumed, still under the leadership of Adolf Florensa and the Service for the Conservation and Restoration of Historic Monuments. This continuity of both the works and the people in charge of them before and after the war was a very rare occurrence in Spain. It was the result, however, of a significant change in the narrative associated with the idealization of Barcelona's Gothic Quarter and its construction: while Catalan nationalism interpreted the quarter as the glory of the medieval city and the evocation of the times when the Catalans dominated trade in the Mediterranean, after the Spanish Civil War the Franco regime considered the quarter to be an expression of the unity of Spanish architecture.

Thus, during the Franco dictatorship, the Gothic Quarter reflected the 'austere and elegant Spanish Gothic style'. This is what may be read in the Falangist weekly *Fotos* from 1942, where it is also stated that

> there was a time when uncivilized separatism sought to see these stones as venerable arguments or levers to move naive well-intentioned hearts. Those were times when the music was Catalan [ … ], sport was political and even the piety had its nuances [ … ], so no one should be surprised that buildings from the distant past were used to stir up regrettable feelings. (Claramunt 15)

The stones and the architecture were the same but the interpretation was almost completely the opposite. In this respect it should be pointed out that it was during the Franco dictatorship that the works surged ahead with the greatest impetus. Indeed, the aerial bombing of Barcelona during the harshest periods of the Spanish Civil War, when the area around the cathedral was severely affected, made it necessary to deal with the problem of reconstruction and to define a dialectic relationship between the pre- and post-war plans and the ruins (Muñoz-Rojas 119–59).[14] Hence, under the Franco regime many buildings were improved and some squares were redefined or even created. This was the case, for example, of Plaça Sant Felip Neri, which was completely remodelled with the reconstruction of the Coppersmith's Guild House (Gremi de Calderers), a building which, as previously mentioned, had been moved to Plaça Lesseps during the demolition works for the Via Laietana, and the case of the rebuilding of the Shoemaker's Guild House (Gremi de Sabaters), which had remained in storage in a municipal warehouse since the time of the demolitions.

Finally, in the 1950s and 1960s, the Gothic Quarter as we know it today was consolidated. Important factors in this consolidation were the guidebooks published about the quarter. These, in actual fact, were manuals of 'artialisation' – to use the term proposed by Alain Roger. The guidebooks made it possible for society and individuals to interact with the invented place and with its narrative. They established the boundaries of the quarter and proposed walks of discovery with maps on which the main buildings were pinpointed and explained together with constant references to the history of Catalonia and Spain. Any reference to the invention of the quarter was avoided at all times and its purportedly 'glorious' past was reiterated, emphasizing the interest of the Roman and Baroque vestiges in the area. Thus, as Florensa himself stated towards the end of his life, tourism was the other driving force in the construction of the Gothic Quarter. Indeed, it was the true mediator with its ritual interaction between visitors – both local and from outside the city – and the Quarter. In this way, tourism has conveyed the value of the quarter to its visitors, while at the same time helping visitors to forget the true origin of the quarter's stones.[15]

## EPILOGUE

Today, after this long and complex process of invention and construction, preservation has overtaken the Gothic Quarter. Despite the regulations imposed by preservation laws and the established scientific approach to interventions, new projects in the area are often based on idealized models similar to those of Florensa. One sees this, for example, on carefully considering the transformation of the façade in the recent refurbishment of an old building at Carrer dels Lledó and its conversion into a hotel by the architect Rafael Moneo. The intervention on the façade is concentrated on the restoration of the arched entrance and of the existing windows' stonework. In this respect, it is at the very least surprising to find that new stonework appears throughout the façade. It should also be pointed out that a design effort has been made to turn the lobby of the hotel into an interior courtyard – like in other Barcelonian medieval Gothic palaces. Comments of this type could also be made with respect to other contemporary projects, like the small new Gothic Hotel at Plaça de Sant Felip Neri by the Barcelona architects Ribas & Ribas. These contemporary examples strengthen the narration defined by the construction of the quarter in the twentieth century, which is closely related to the touristic exploitation of the city, while ignoring the dissonances in relation to lifestyles and neighbourhoods that have arisen in this process. In the years when Florensa was building Barcelona's Gothic Quarter, the Barcelona poet and art historian Juan Eduardo Cirlot wrote a book entitled *Ferias y atracciones* (Amusement Parks and Funfair Attractions) as a volume in the series called *Esto es España* (This is Spain). Here he presents funfairs and amusement parks as things that define what we are, that define our identity and our heritage. That is precisely what the Gothic Quarter is like. The parallelism is enlightening. Both funfair attractions and amusement parks, on the one hand, and the Gothic Quarter, on the other, alter the routines of their regular or occasional visitors and create the sensation of making an imaginary journey. Both are delimited spaces full of attractions where prizes or souvenirs are to be had. Both of them are places that require the mediation of a map that defines their limits and shows the main paths and routes for discovering their main points of interest and that provides information on them. Likewise, in both of them music is essential in defining their atmosphere – the unmistakable melody of the attractions in an amusement park and the street musicians in the Gothic Quarter. Further parallelisms could be added but here we will just mention one more: light, or more precisely, artificial lighting (electric or neon). Artificial lighting brings out funfair buildings and attractions with its magically ravishing glow, and it highlights the features of the Gothic Quarter and many other 'monuments', as it has ever since its invention. This is the same light that George Steiner would consider revealing of the Gothic Quarter's artificial construction, just as it reveals the artificial construction of Warsaw and of many other sites; it is a lighting that has the artificial aftertaste of neon (Steiner 26). In this respect it should perhaps not be forgotten that 'neon' comes from the Latin *neos* – meaning new, and moreover neon is a noble gas. In short, it produces a 'new and noble' light.

The construction of Barcelona's Gothic Quarter shows how heritage sites, just like cities and their historic urban fabric, are created and constantly reshaped by an ongoing Bourdieuian struggle and negotiation between different social, cultural and political groups seeking to control their symbolic and economic capital. It is a struggle that defines and spreads narrations and imaginaries through the use of devices such as newspapers, literature, architecture and tourism, creating and negotiating through these devices the imaginary and symbolism of cities and heritage sites as well as the illusion of their authenticity. In short, just as is shown by Barcelona's Gothic Quarter, authenticity is based on social, political and cultural struggle rather than on material factors.

## BIBLIOGRAPHY

Barenys, José M. 'Luis Domènech y Muntaner'. *Urbanización y Edificaciones. Revista Técnica* 3 (1924): 8–11. Print.

Buls, Charles. *Esthétique des Villes*. Brussels: Imprimerie Bruylant-Christophe, 1893. Print.

Camille, Michel. *The Gargoyles of Notre Dame*. Chicago: University of Chicago Press, 2009. Print.

Cirlot, Juan Eduardo. *Ferias y Atracciones*. Barcelona: Editorial Argos, 1950. Print.

Claramunt, Jorge. 'En el Corazón de Barcelona. Vigencia y Actualidad de su Maravilloso Barrio Gótico'. *Fotos* 262 (1942): 15–20. Print.

Cócola Gant, Agustín. *El Barrio Gótico de Barcelona. Planificación del Pasado e Imagen de Marca*. Barcelona: Madroño, 2011. Print.

——. 'The Invention of the Barcelona Gothic Quarter'. *Journal of Heritage Tourism* 9.1 (2014): 18–34. Print.

——. 'El Barrio Gótico de Barcelona. De Símbolo Nacional a Parquet Temático'. *Scripta Nova. Revista Electrónica de Geografía y Ciencias Sociales* XV.371 (2011). Universitat de Barcelona. Web. 17 Jan. 2015.

Cosgrove, Denis. 'Heritage and History: A Venetian Geography Lesson'. *Rethinking Heritage: Culture and Politics in Europe*. Ed. Shannan Peckham. London: I.B. Tauris, 2003. 113–23. Print.

Cuccu, Marina. 'Las Calles de Barcelona de Víctor Balaguer'. *Barcelona Quaderns d'Història* 14 (2008): 147–61. Print.

Domènech i Montaner, Lluís. 'Reforma de Barcelona'. *La Renaixensa* 3 (1879): 125–48. Print.

Domènech i Girbau, Lluís. 'La Reforma de Barcelona'. *Lluís Domènech i Montaner, Arquitecte Modernista*. Barcelona: Fundació Caixa Barcelona, 1989. 217–19. Print.

Durán Sanpere, Agustín. *El Barrio Gótico y su Historia*. Barcelona: Aymà, 1962. Print.

Enlart, Camille. *Manuel d'Archéologie Française, Depuis les Temps Reculés Jusqu'à la Renaissance. TI, TII, TIII*. Paris: Picard, 1902–1916. Print.

Florensa, Adolfo. *Nombre, Extensión y Política del 'Barrio Gótico'*. Barcelona: Ayuntamiento de Barcelona, 1958. Print.

——. *La Plaza de San Felipe Neri Ayer, Hoy y Mañana*. Barcelona: Ayuntamiento de Barcelona, 1958. Print.

——. 'Restauraciones y Excavaciones en Barcelona Durante los Ultimos Veinticinco Años'. *Cuadernos de Arqueología e Historia de la Ciudad* 6 (1964). 5–36. Print.

——. *Conservación y Restauración de Monumentos Históricos (1927–62)*. Barcelona: Ayuntamiento de Barcelona, 1967. Print.

Ganau, Joan. *Els Inicis del Pensament Conservacionista en l'Urbanisme Català, 1844–1931*. Barcelona: Publicacions de l'Abadia de Montserrat, 1997. Print.

——. 'La Ciutat com a Museu: Les Obres de Reforma Interior i el Naixement del Barri Gòtic de Barcelona (1907–1930)'. *Expansió Urbana i Planejament a Barcelona*. Ed. Joan Roca. Barcelona: Institut Municipal d'Història – Edicions Proa, 1997. 193–205. Print.

——. 'La Recreació del Passat: el Barri Gòtic de Barcelona, 1880–1950'. *Barcelona Quaderns d'Història* 8 (2003): 257–72. Print.

——. 'Invention and Authenticity in Barcelona's "Barri Gòtic"'. *Future Anterior* 3.2 (2006): 11–23. Print.

——. 'Reinventing Memories: The Origin and Development of Barcelona's Barri Gòtic, 1880–1950'. *Journal of Urban History* 34 (2008): 795–832. Print.

Garcia-Fuentes, Josep-Maria. 'On the Creative Interpretation of the Past: Barcelona's *Barri Gòtic*'. *Chronocity-2010. Sensitive Interventions in Historic Environment*. Ed. Dimitra Babalis. Florence: University of Florence and School of Architecture of Athens, 2010. Print.

——. 'Víctor Balaguer and the Catalan-Spanish Monasteries. On the Invention of Heritage, Monuments and Traditions'. *Future Anterior* 10.1 (2013): 40–51. Print.

Giralt, Ricard. 'Urbanización del Barrio Barroco de la Virreina'. *La Construcción* 11 (1917): 12–14. Print.

Guàrdia, Manel and Josep-Maria Garcia-Fuentes. 'La Construcció de l'Eixample'. *L'Eixample: Gènesi i Construcció*. Barcelona: Lunwerg, 2009. 47–95. Print.

Junyent i Rafat, Josep. *Jaume Colell i Bancells: Les Campanyes Patriòtico-religioses: 1878–1888*. Vic: Patronat d'Estudis Ausonencs, 1990. Print.

Llompart, Amadeo. 'El urbanismo en la Escuela de Barcelona'. *Arquitectura* 71 (1925). 45–46. Print.

Lowenthal, David. *The Past is a Foreign Country*. Cambridge: Cambridge University Press, 1985. Print.

Muñoz-Rojas, Olivia. *Ashes and Granite. Destruction and Reconstruction in the Spanish Civil War and Its Aftermath*. Brighton: Sussex Academic Press and Catalan Observatory LSE, 2011. Print.

Murphy, Kevin D. *Memory and Modernity. Viollet-le-Duc at Vézelay*. Pennsylvania: The Pennsylvania State University Press, 2000. Print.

Nicolau, Antoni and Daniel Venteo. 'La Monumentalització del Centre Històric: La Invenció del Barri Gòtic'. *La Construcció de la Gran Barcelona: l'Obertura de la Via Laietana, 1908–1958*. Barcelona: Ajuntament de Barcelona, 2001. 100–127. Print.

Pieró, Xavier. 'Adolf Florensa i el Patrimoni Arquitectònic de la Ciutat de Barcelona. La Seva Labor en la Restauració de Monuments i Conjunts'. *Adolf Florensa i Ferrer (1889–1968)*. Barcelona: Ajuntament de Barcelona, 2002. 33–88. Print.

Puig i Cadafalch, Josep. 'La Casa Catalana'. *Congrés d'Història de la Corona d'Aragó dedicat al Rey en Jaume I y a la Seua Epoca. Segona Part*. Barcelona: Estampa d'en Francisco Altés, 1913. 1041. Print.

——. 'La Destrucció de l'Acròpolis de Barcelona' *La Veu de Catalunya* 10 Dec. 1924. 7. Print.

Roger, Alain. *Court Traité du Paysage*. Paris: Éditions Gallimard, 1997. Print.

Rubió i Bellver, Joan. *Taber Mons Barcenonensis*. Barcelona: Casa de la Caritat, 1927. Print.

Rucabado, Ramon. 'Un Barrio Gótico en Barcelona'. *La Cataluña* 189 (1911): 305–11. Print.

Sala, Carmelo. *El Barrio Gótico: sus Calles, sus Gentes, su Vida*. Barcelona: [s.n.], 1976. Print.

——. *El Barrio Gótico: Su Historia, su Intimidad*. Barcelona: Talleres de Negre, 1977. Print.

Sitte, Camillo. *Der Städtebau nach Seinen Künstlerischen Grundsätzen*. Vienna: Graeme, 1889. Print.

Solà-Morales, Ignasi de. *Joan Rubió y Bellver y la Fortuna del Gaudinismo*. Barcelona: Colegio Oficial de Arquitectos de Cataluña y Baleares, 1975. Print.

Steiner, George *The Idea of Europe*. Tilburg: Nexus Institute, 2004. Print.

Street, George Edmund. *Some Account of Gothic Architecture in Spain*. London: J. Murray, 1865. Print.

Torrella, Rafel. 'La Fotografia al Concurs Artístic de la vella Barcelona'. *La Construcció de la Gran Barcelona: l'Obertura de la Via Laietana, 1908–1958*. Barcelona: Ajuntament de Barcelona, 2001. 128–47. Print.

Verdaguer, Jacint. Á Barcelona: Oda. Barcelona: Estampa Espanyola, 1883. Print.

Viollet-le-Duc, Eugène-Emmanuel. *Dictionnaire de l'Architecture Française du XI<sup>e</sup> au XVI<sup>e</sup> siècle*. Paris: A. Morel Editeur, 1854–1868. Print.

## NOTES

1   The geographer Joan Ganau (1997; 2003) was the first scholar to thoughtfully study the construction of Barcelona's Gothic Quarter.

2   For one account of this struggle and a possible political, social and cultural use of the cathedral's new façade by the Catalan Catholic Church, see Junyent i Rafart (123–6) and Ganau (1997 199–240).

3   Cerdà's project was imposed by the government in Madrid because they were in charge and had a different criteria than the municipal government – who organized the competition without Madrid's collaboration. Cerdàs project was the best proposal according to Madrid's criteria.

4   'Sant Jordi de l'Audiencia vol veure Santa Clara; l'antich Palau dels Comtes anyora'l del Concell. ¡Oh! atèrra eixa cortina de cases que separa l'estátua de Don Jaume del seu real Tinell. En mitx d'aqueixa plassa, que no tindrá segona, les tres Columnes d'Hèrcules quan mire'l viatjer, creurá veure les Gracies per ferte de corona, de brassos enllassades, dansant en ton verger' (translation by author).

5   The Via A plotted by Cerdà in his plan and built between 1908 and 1913 is the avenue known today as Via Laietana.

6   Indeed, one of the main challenges during the early years of the construction of Barcelona's new expansion area – the Eixample – was to make it appealing so that the city's elites and upper social classes would move there. These groups felt that the new expansion district did not meet their wishes and aspirations for social representation, as is revealed, for example, by Eusebi Güell's decision in the late nineteenth century to build his new palace in the Old City and not in the new Eixample. In this respect, the most successful initiative to make the new expansion appropriate for these people was the naming of the streets by Victor Balaguer in 1863 (Cuccu 147–61; Garcia-Fuentes 2013 43). The relocation of ancient buildings from the Old City should be seen as part of this strategy too. However, the elites' misgivings about the new expansion area lasted until the Modernista generation succeeded in bringing about a significant change in the district's imaginary (Guàrdia and Garcia-Fuentes 47–95).

7   This is the area where the Rambla del Raval is now found.

8   Joan Rubió i Bellver was a disciple of Gaudí (Solà-Morales).

9   *El Imán. Revista enciclopédica; ilustrada; mensual* 58 (1927).

10   This title referred to the Roman name of Barcelona's ancient Acropolis.

11   The art historian Agustín Cócola Gant is the author of what is probably the most accurate study of the Gothic Quarter's historical construction. His PhD dissertation at the University of Barcelona was specifically devoted to this subject and was published as a book in Spanish (*El Barrio Gótico de Barcelona. Planificación del Pasado e Imagen de Marca*).

12   Regarding the debate about the 'Catalan house' and the construction of Barcelona's Gothic Quarter, see Agustín Cócola (2011).

13   In connection with chimeras, nineteenth-century Gothic and Viollet-le-Duc, see the brilliant book by Michael Camille, *The Gargoyles of Notre Dame*.

14    For an accurate account of the post-war reconstruction in Barcelona and its
      relationship to the situations in Madrid and Bilbao and to the contemporary Spanish
      context, see the extensive and precise book by Olivia Muñoz-Rojas, *Ashes and Granite.*
      *Destruction and Reconstruction in the Spanish Civil War and Its Aftermath*.

15    In the 1970s, dissonances concerning lifestyles and neighbourhoods arose with
      the increased touristic exploitation of the quarter. An early witness to this process
      is the somewhat elegiac book, *El Barrio Gótico: sus Calles, sus Gentes, su Vida*, by
      Carmelo Sala on the local intimate atmosphere of the quarter and its progressive
      contemporary transformation.

# 3

# Building Reconstructions and History Constructions in Hungary and Romania under Communist Rule

*Robert Born*

In post-1945 Communist Eastern Europe, monument preservation served various purposes, which are analysed in this comparative study addressing reconstruction projects in Romania and Hungary. This chapter focuses on the period from the 1950s onwards and examines the agendas underlying a number of projects in terms of the two countries' policies of commemoration. In the case of Hungary, efforts concentrated on architectural remnants in Budapest and Esztergom distinguished by their associations with the Anjou and Corvinus royal dynasties of the late Middle Ages and the Renaissance. Conversely, Romania sought to reinforce the alleged roots of the nation in classical antiquity by means of ambitious reconstructions such as the triumphal monument in Adamclisi originally built by Emperor Trajan. Apart from attributing these projects to endeavours to forge a national identity, the architectural initiatives are discussed regarding the specific circumstances under which heritage conservators had to operate in Romania and Hungary.

The end of the Second World War and the imposition of the people's democracies in Central and Eastern Europe just a few years later had far-reaching repercussions for the treatment of the architectural heritage in this large region. In this chapter, an important complex of heritage conservation in Central Europe is examined using the example of reconstruction projects in Hungary and Romania. Initially, the conservation authorities in these two neighbouring countries operated under similar conditions. State institutions entrusted with the excavation, study and preservation of their historical heritage had been in existence in both countries since the second half of the nineteenth century.[1] In addition, the circumstances surrounding their work were similar owing to the nationalization of land ownership in the aftermath of the Communist takeover in the two countries (Entz and Gerő). In the initial period after this regime change in Romania and Hungary (as in the other countries in East Central Europe), heritage conservation policy was influenced by ideological guidelines from the USSR. Hungary's orientation to Soviet legislation and the institutional affiliation of heritage conservation to the Academy of Sciences in Romania in 1951 are symptomatic of this realignment.[2] Despite these structural similarities, differences emerged regarding conservation practice and above all the reconstruction of historical buildings among the Eastern Bloc countries. These differences can be explained by the different degrees of destruction suffered during the war and the deliberate destruction of historical monuments and other cultural possessions by the occupying German forces. Consequently, the possibility of reconstruction schemes was mainly raised by the state conservation authorities in those countries where the losses of monuments and listed buildings had been especially severe and widespread, such as in the Soviet Union, Poland and Hungary.

In Hungary, the two regional priorities earmarked for reconstruction schemes were Budapest and Sopron. Fighting over Sopron in north-west Hungary had been especially fierce because the leadership of the Arrow Cross Party, which had come to power in October 1944 with the support of Nazi Germany, had taken refuge there. Meanwhile in the Hungarian capital, above all large areas of Castle Hill in Buda had been destroyed in the fighting (Dercsényi 99–103). The destruction of extensive built-up areas provided an opportunity for large-scale archaeological excavations. During this research, important structures of the royal capital of the Hungarian kings in the late Middle Ages and the early Renaissance were unearthed in the area of the former Royal Palace. The discovery of architectural elements enabled the reconstruction of the Gothic Hall using the process of anastylosis (Gerő).

The strategy applied on Castle Hill in Budapest of studying stone architectural elements followed by anastylosis and reconstruction became common practice in Hungarian conservation as of the 1950s. Other prominent examples of this approach include the palace chapel in Esztergom and the late Gothic canopy fountain in Visegrád (Marosi 2001 273–6). These reconstructions were primarily intended to enhance the interior rooms and harked back to strategies from the interwar period. This conceptual continuity was maintained by a number of specialists who remained senior conservation officials despite the new regime. Alongside architects László Gerő (1909–95) and Kálmán Lux (1880–1961), they included Tibor Gerevich (1882–1954), an especially influential art historian and cultural politician in the interwar period (Szűcs).

Similar personnel and conceptual continuity is also noticeable in connection with the reconstruction of the former Royal Palace on Castle Hill in Buda, which took place in the direct vicinity of the aforementioned archaeological excavation and reconstruction work. This was, however, a project with heavy ideological baggage. In contrast to the reconstruction of the Gothic Hall in Buda and the palace chapel in Esztergom, which were seen as testimony to the exquisite cultural prosperity of the Kingdom of Hungary in the late Middle Ages and the early Renaissance, the Royal Palace was an architectural ensemble which had been instigated by the Habsburgs. In addition, the sprawling complex had been used by Hungary's autocratic ruler Miklós Hothy (1868–1957) as his residence in the interwar period. Even so, shortly after seizing power, in autumn 1949 the Communist rulers decided to turn the palace (which at this time still lay in ruins) into the headquarters of the party and seat of government. Its isolated yet dominant position over the city held out the prospect of effective control coupled with security should troubled times arise. The restoration of the palace was enshrined as an important objective in the first five-year plan.

Party leader Mátyás Rákosi (1892–1971) supported the restoration of the complex with particular attention being paid to the central dome designed by architect Alois Hauszmann (1847–1926), which had originally only been built when the complex was completed between 1891 and 1905. The reconstruction of the palace was finalized with the completion of the dome designed by architect Lajos Hidasi. Although partly reminiscent of Roman Baroque domes, the simplicity of the modern dome's design has clear echoes of the historicist style of architecture of the Stalin era (Figure 3.1). Immediately after completion, the new regime under János Kádár (1912–89) called for Buda Castle to be transformed into a centre of national culture. This decision prompted the extensive redevelopment of the palace's interior, which dragged on for a number of decades until the end of the Communist regime in Hungary (Rostás 2011). With Buda Castle dominating Budapest's skyline, its size was not the only factor making it unique among the reconstruction projects in Communist Hungary. As of the mid-1960s, reconstructions were increasingly determined by the principles set out in the Charter of Venice (1964), with anastylosis becoming the main concept employed.[3]

A completely opposite trend emerged at about the same time in neighbouring Romania, where conservation was shaped by the doctrine of National Communism adopted by Nicolae Ceauşescu (1918–89). After a period of rapprochement with the West, in 1971 a radical about-turn was announced by Ceauşescu in his 'July Theses' inspired by a trip to North Korea, China and North Vietnam. Mapping out the country's route to cultural autochthonism and amounting to a miniature cultural revolution, the 17 proposals of his new programme called for Romania to resist external cosmopolitan influences and for the creation of a 'new man' (Verdery 101, 107–8). This policy naturally encompassed historiography and the teaching of history. Both areas were assigned a key ideological role, as became starkly apparent with the historical introduction, over 30 pages in length, to the 1974 party programme entitled 'Creating a Multilaterally Developed Socialist Society and Romania's Path towards Communism'. The conceptual elements it emphasised such as the ancient origins of the Romanian nation, its continuity on Romanian territory since ancient times, its constant unity and the relentless struggle to defend its independence formed the template for all accounts of history written until 1989 (Petrescu). The special importance of Romania's history policy in its self-portrayal in those years was brought home by the National Museum of Romanian History (Muzeul Naţional de Istorie a României), opened in 1972 in Bucharest, which was assigned an important role in the 'patriotic education of the masses'. The centrepiece of the museum was the plaster casts of Trajan's Column, which was described in the press as the 'birth certificate of the Romanian people'.[4] Both the decision to set up this museum and the eventual delivery of the casts (originally commissioned under the monarchy, they had been held up in Rome by the turmoil of the Second World War) underline the radical change of policy in Romania during those years (Simionescu and Padiou 220).

In addition to Trajan's Column, the ancient Tropaeum Traiani – a victory monument built in the district of Adamclisi near the Lower Danube – was one of the most important historical symbols of Romanian national identity.

3.1    Budapest: The reconstructed royal palace with the dome designed by the architect Lajos Hidasi
*Source:* Robert Born.

Accordingly, this triumphal monument erected by order of Emperor Trajan together with a monumental altar in 108/9 AD was repeatedly excavated following its rediscovery in the second half of the nineteenth century – partly because by this time it had badly decayed, although the cylindrical core made out of cast mortar as well as a considerable number of the bas-reliefs had survived. Ornamental tendrils and metopes were found scattered around the base (one had also been discovered to have been reused as part of a fountain) along with the tropaion (formerly the uppermost part) (Florescu). For many years, intense debate raged among scholars about the precise original arrangement of these elements and the exact proportions of the base and the upper part containing captured weapons. Nevertheless, politicians of all colour were determined to see the Tropaeum Traiani rebuilt in the twentieth century. Following an act of Parliament in 1890, the original parts of the Roman triumphal monument were moved from Adamclisi to Bucharest, where they were to primarily serve educational purposes (Florescu 26f). Since the implementation of this project was delayed, the metopes from the Tropaeum Traiani were displayed on the terrace of the Military Museum, a building originally erected in the New Romanian style as an art palace for the National Exhibition in 1906. When the government decided to create a memorial to the unknown soldier in 1923, the location chosen for it was also the terrace of the Military Museum. The inscription emphasized soldiers' sacrifice for the restoration of the territorial unity of Romania (Tuca and Gheorghe 43–4). The originally universal semantics of the cult of the unknown soldier was thus transformed into a central symbolic place of the nation state.[5] Although this shift in emphasis is doubtless not a genuinely Romanian development, the assignment of this symbol of national unity to a line of heritage starting with Emperor Trajan's military triumph (symbolized by the original fragments from the Tropaeum Traiani) is nothing short of extraordinary. This idea was taken a step further by a plan drawn up in 1943 to reconstruct the Tropaeum Traiani as a mausoleum in Bucharest for soldiers who had fallen during the ongoing war, although it did not get beyond the competition phase.[6] The only comparable parallel is the Tomb of the Unknown Soldier in Rome, which was built in 1921 in the *Altare della Patria* – the 'Altar of the Fatherland'. Due to its location on the Capitoline Hill beneath the statue of goddess Roma, it laid claim to the legacy of ancient Rome by the then young Italian nation state (Inglis 14).

After the seizure of power by the Communists in Romania in 1947, the sculptures from the Tropaeum Traiani initially remained in the courtyard of Bucharest Museum of Antiquities. A change in tack only became apparent in 1973, when a panel of experts from the party leadership was charged with developing a solution for the reconstruction of the victory monument at its original location. The design produced by architect Dan Rusovan called for the original core of the monument to be surrounded by a self-supporting steel structure bearing casts of the surviving original pieces and the missing elements (Figure 3.2). Only the steps in the base were built out of solid stone from a nearby quarry (like the ancient workpieces). A brand new museum was built in Adamclisi to preserve the original pieces. Construction work on the two complexes began in autumn 1975.[7]

3.2  Adamclisi (Romania): The original Roman core of the triumphal monument surrounded by a self-supporting steel structure bearing casts of the surviving original pieces and the missing elements *Source:* Photography from Rădulescu, Figure 3 (n.p.): https://revistaponticaen.files.wordpress.com/2012/07/pontica-10-1977.pdf.

The inauguration was held on 28 May 1977 in connection with a working visit by party leader Nicolae Ceaușescu to the Constanța district.[8] During the ceremony, soldiers holding state and party flags were posted on the roof of the reconstructed edifice (Figure 3.3). This elaborate spectacle took place shortly after the celebrations marking the centenary of Romania's independence on 9 May 1977.

In his speech at this event, Ceaușescu drew a link between the settlement's continuity and Romanian independence. Mentioning the military defeat of the Dacians as a precursor of the development of the Romanian people, in the same breath he also stressed the continuation of typical Dacian virtues such as 'the unquenchable thirst for freedom, the refusal to bend to a foreign yoke, the determination to remain true to oneself and to shape one's own life and destiny alone'.[9] Given this construction of history, the combination of the inauguration of the rebuilt Tropaeum Traiani with the launch of the first Romanian bulk tanker named Independenţa ('Independence') and a visit to the construction site of the Danube–Black Sea Canal as part of the leadership's working visit acquired new symbolism. A spiritual bond was forged between the prestige projects of a Communist Romania striving for economic self-sufficiency and the ancient monuments (Barnea and Panaite 226). The completion of the reconstruction of Trajan's victory monument in Adamclisi was an important milestone in the Ceaușescu regime's policies of National Communism. It is also important to note that the execution of this project coincided with the dissolution of the state conservation institutions by order of the Communist Party leadership. Consequently, listed buildings and other historic monuments in Romania suffered large-scale destruction as a result of forced industrialization and Ceaușescu's megalomaniac plans to redevelop Bucharest (Giurescu).

The political symbolism of the reconstruction of the Tropaeum Traiani survived the collapse of Communism in Romania in 1989 unscathed. Over the past two decades, images of it have repeatedly provided a backdrop to ceremonies involving the country's leadership and election campaign events (Barnea and Panaite 228–30).

3.3    Adamclisi (Romania): The inauguration ceremony for the reconstructed triumphal monument on 28 May 1977 with the delegation around party leader Nicolae Ceauşescu *Source:* Photo # LA482, Online Communist Photo Collection, *(12.12.2014).*

To sum up, the reconstruction activities under the Communist authorities in both Hungary and Romania show patterns of continuity with the interwar period and even the dawn of the twentieth century. In the case of Hungary, this was due to the ongoing presence of leading specialists in conservation authorities. Similar circumstances have also been noted regarding reconstruction projects in Communist Poland (Majewski 16–19). In the case of Romania, however, ideological continuities can be observed which can doubtless be interpreted as a radical approach to dealing with its monument heritage compared to the rest of the Communist bloc.

## BIBLIOGRAPHY

Barcza, Géza. 'Szocialista Országok Műemléki Törvényei'. *Magyar Műemlékvédelem.* Vol. 2 (1959–60): 19–30. Print.

Barnea, Alexandru and Adriana Panaite. 'Tropaeum Traiani. Monument şi Propaganda'. *Caietele* ARA 1 (2010): 223–34. Print.

Bartoldy, István, and Andrea Haris, eds. *A Magyar Műemlékvédelem Korszakai. Tanulmányok.* Budapest: Országos Műemlékvédelmi Hivatal, 1996 (Művészettörténet – Műemlékvédelem 9). Print.

Dercsényi, Dezső. 'La Tutela dei Monumenti in Ungheria Dopo la Liberazione'. *Acta Historiae Artium* 2 (1955): 97–132. Print.

Entz, Géza and László Gerő. 'Műemlékvédelmi Tapasztalatok Szocialista Országokban'. *Magyar Műemlékvédelem.* Vol. 2 (1959–60): 13–18. Print.

Entz, Géza, ed. *A Velencei Carta és a Magyar Müemlékvédelem.* Budapest: [s.n.], 1967. Print.

Florescu, Bobu Floarea. *Das Siegesdenkmal von Adamklissi, Tropaeum Traiani*. Bucharest-Bonn: Verlag der Akademie der Rumänischen Volksrepublik, 1965. Print.

Gerő, László. 'Budai Középkori Királyi vár és vár Maradványainak Helyreállítása'. *Magyar Műemlékvédelem*. Vol. 1 (1967/68): 155–94. Print.

Giurescu, Dinu C. *The Razing of Romania's Past: International Preservation Report*. Washington, D.C: U.S. Committee, International Council on Monuments and Sites, 1989. Print.

Inglis, K. S. 'Unknown Soldiers: From London and Paris to Baghdad'. *History & Memory. Studies in Representation of the Past* 5.2 (1993): 7–31. Print.

Ioan, Augustin. 'Qu'est-ce que c'est que l'Architecture "au Spécifique National" et Comment se Révèle-t-elle?' *Nation and National Ideology: Past, Present and Prospects : Proceedings of the International Symposium Held at the New Europe College, Bucharest, April 6–7, 2001*. Bucharest: New Europe College, 2002. 288–305. Print.

Kosseleck, Reinhart. 'Der Unbekannte Soldat als Nationalsymbol im Blick auf Reiterdenkmale'. *Vorträge aus dem Warburg-Haus* 7 (2003): 137–66. Print.

Majewski, Piotr. *Ideologia i Konserwacja. Architektura Zabytkowa w Polsce w Czasach Socrealizmu*. Warszawa: Wydawn. Trio, 2009. Print.

Marosi, Ernő. 'Die Anfänge der Denkmalpflege in Ungarn und die Tätigkeit der k. u. k. Zentralkommission in Ungarn'. *Die Ungarische Kunstgeschichte und die Wiener Schule 1846–1930. Catalogue of the Exhibition Collegium Hungaricum Vienna 1983*. Ed. Marosi, Ernő and Gizella Cennerné Wilhelmb. Budapest: Statistischer Verlag, 1983. 13–18. Print.

——. 'Drei Mittelalterliche Schlüsseldenkmälern der Kunstgeschichte Ungarns-Restauriert. Székesfehérvár, Esztergom und Visegrád im Jahr 2000'. *Acta Historiae Artium* 42 (2001): 255–81. Print.

Opriş, Mihai. *Comisiunea Monumentelor Istorice*. Bucharest: Editura Enciclopedică, 1994. Print.

Petrescu, Dragoş. 'Historical Myths, Legitimating Discourses, and Identity Politics in Ceauşescu's Romania (Part 1)'. *Radio Free Europe/Radio Liberty East European Perspectives* 6.7 (2004). Web. 10 June 2015. <http://www.rferl.org/content/article/1342455.html>.

Rădulescu, Adrian. 'Un act de Semnificaţie Patriotică: Reconstruierea Monumentului Tiumfal de la Adamclisi'. *Pontica* 10 (1977): 9–14. Print.

Rostás, Péter. 'A Nemzet Palotája. A Budavári Palota elmúlt évszázada'. *Magyar Narancs* 23/15 (2011): 31–3. Print.

Rusovan, Dan. 'Conservarea şi Valorificarea Muzeistică a Monumentului Tiumfal Tropaeum Traiani'. *Pontica* 10 (1977): 15–20. Print.

Simionescu, Paul and Hubert Padiou. 'Comment le Musée National de Bucarest Racontait l'Histoire'. *A l'Est, la Mémoire Retrouvée*. Eds Alain Broissant, et al. Paris: La Découverte, 1990. 212–28. Print.

Szűcs, Julianna, P. 'A Művészettörténész Mint Kultúrpolitikus. Gerevich Tibor Vonzásai és Választásai'. *Kritika* 2 (1997): 27–9. Print.

Tuca, Florian and Cristache Gheorghe. *Altarele Eroilor Neamului. Monumente şi Însemne Memoriale în Aria de Träire Româneasca*. Bucharest : Editura Europa Nova, 1994. Print.

Verdery, Katherine. *National Ideology under Socialism: Identity and Cultural Politics in Ceauşescu's Romania*. Berkeley: University of California Press, 1991. (*Societies and Culture in East-Central Europe 7*). Print.

Zach, Krista: 'Von Burebista bis Ceaușescu. Der Mythos vom Zweitausendjährigen
    "Unabhängigen Einheitsstaat"'. *Wissenschaftlicher Dienst Südosteuropa* 28 (1979):
    200–205. Print.

## NOTES

1    For more on Hungary, see Marosi 1983; Bartoldy and Haris; for more on Romania,
     see Opriș.

2    For more on this, see Barcza; for more on Romania, see Opriș 161.

3    Regarding the influence of the principles of the Charter of Venice on conservation in
     Hungary, see Entz 1967.

4    See the report in *Scînteia* 9142.41 dated 9 May 1972.

5    See Kosseleck.

6    For more on the competition, see Ioan 293.

7    Rădulescu; Rusovan.

8    See the report in: *Scînteia* 10815.46 dated 29 May 1977.

9    Ceaușescu, Nicolae. Speech marking the centenary of independence. Quoted from
     Zach 202.

# Rebuilding the Jesuit Church and College: The Power of Toponymy, Representations and Catholic Advocacy in Nineteenth–Twentieth-Century São Paulo, Brazil

*Renato Cymbalista and João Carlos Santos Kuhn*

## INTRODUCTION

The city of São Paulo, Brazil's largest metropolis, the capital of the state of the same name and the most important capital in the country, consecrated, as part of its history, a precise point that symbolizes its place of birth: the Jesuit church, erected on top of the hill dividing the basins of the Tamanduateí and Anhangabaú rivers. The year was 1554 and the date – which was surely carefully chosen by the Jesuit – was 25 January, São Paulo's Day, in honour of the first apostle to preach to the Gentiles.

From that centre, the Jesuits conducted their catechetical endeavours across the region, and the city began to structure itself. Fifty kilometres away from the Atlantic coast, São Paulo provided waterway access to a vast hinterland, while also serving as a gateway for exploring and occupying a large part of the Brazilian inlands.

The first Jesuit lodge in São Paulo was described by priest José de Anchieta, through the letters he sent to the superiors of the Society of Jesus, as a *paupercula domo* (humble house), measuring ten by fourteen *craveiros*,[1] where the Jesuit school, infirmary, dormitory, kitchen and pantries were located (Salgado 72). Two years after Anchieta's statement, this first lodge was replaced with a construction made of *taipa de pilão*,[2] and beside it, the *colégio* (college) was built, which was a mix of residences, Jesuit practice areas, and what was the city's main educational institution for a long time. In 1640, the priests were expelled from the city during a conflict with the locals, and their return was not authorized until 13 years later. To celebrate their return to São Paulo, they built a new church on the same site, a more monumental structure made of *taipa de pilão* and stone. They occupied that site until the mid-eighteenth century.

In 1759, the Jesuits once again left the city after the Portuguese crown decreed their expulsion from all of its dominions, the first in a series of similar measures in the whole world that would result in the suppression of the Order for over half a century. The Palace of the Governors was then established in the *colégio* building, taking advantage of one the most imposing building in the city, and the square was renamed as *Largo do Palácio* (Palace Square). The church remained in operation and managed by the Diocese of São Paulo, under the name of *Igreja do Bom Jesus* (Church of Good Lord Jesus), until the end of the nineteenth century.

The first images available date back to the early nineteenth century, when the group of buildings was already being used as the Palace of the Governors and Bom Jesus Church. The first record is from 1818, painted by Thomas Ender, and depicts a set of buildings that had changed little since Jesuit times. In 1822, Brazil gained its independence from Portugal, thus becoming the only monarchy in the Americas. In 1862, the first photographic records of the buildings, coming from Militão de Azevedo, show the same structure from Jesuit times, except for the expansion of the side wing of the *colégio*, indicating how slowly São Paulo modernized during the first half of the nineteenth century.

In the following decades, with the acceleration of urban transformation processes.[3] *Largo do Palácio* surpassed, to a certain degree, its colonial features: the L-shaped unit disappeared with the demolition of one of the *colégio* wings, leaving the Palace of the Governors isolated in the centre of a block, giving it a certain monumentality. The *terreiro*[4] was landscaped and fenced, and the building gained a neo-classical pediment.

Nonetheless, the presence of the church attached to the Palace on the left side of the main façade marked the obstacles in overcoming the colonial past of the monarchic society,[5] which had not yet faced the dilemmas of modernization of Brazilian society under the monarchy, such as an incomplete separation between Church and State and the permanence of slavery.

In 1888, slavery was abolished, and in 1889 the Republican regime was established in Brazil, bringing with it a strong secular drive that separated civil and religious powers, giving to the State the power over the management of the dead, and oriented the major cities towards a general modernization movement, with an eye towards Europe, especially France. From the urban point of view, new times meant many changes, including the battle against a territory strongly structured by religious buildings and its replacement with civilian buildings. In São Paulo, the fate of the Palace/former Jesuit church was sealed in this process. It was unthinkable for the authorities of the time to allow the coexistence of the seat of provincial power with a religious building whose memory referenced the Jesuits, who were considered by the new political elites as the symbol of a colonial and obscurantist past.

The Jesuit church had fallen into disuse, and on a stormy night in 1896, the roof of the old and already ruined Jesuit church in São Paulo collapsed. This event leveraged a debate already underway regarding the future use of that part of the city, its main symbolic landmark. The church's structure was not condemned, but a decision was made to demolish it.

In all likelihood, the church's end was probably already near. Since 1881, Ramos de Azevedo, the period's chief city engineer, was working on transforming *Largo do Palácio* into a secular civic centre. The Largo vicinity was also home to the main administrative buildings of the time: besides the Palace, the State Department, the Department of Finance and Department of Central Policy were built in the last decades of the nineteenth century. In the church's old site next to the Palace,

the square was enlarged and the headquarters of the state legislative power erected, consisting of a set of administrative buildings and an L-shaped plaza that formed a square and a *piazzetta*, thereby creating the capital's first urban set of buildings and the stage for the city's civic protests in the very same site (Campos, C.8).

4.1    Palace of the Governors in the early 1860s, keeping the Jesuit church and college features *Source:* Militão Augusto de Azevedo. Créditos Fotográficos das reproduções: Hélio Nobre/ José Rosal.

In the eyes of the Republican elite, the territory was stabilized with the end of the church and construction of the administrative compound: the process of secularizing such a significant place on the historic hill where the city was founded was complete. However, the history of the following decades reveals the remarkable resilience of the site's traditional meanings, its resistance to secularization and the site's definition to an erudite area with balanced proportions and architecture. As everyone knows, the centre of contemporary São Paulo is not home to the Republican and secular urban set, but rather the rebuilt church and Jesuit College.

The backlash began with the demolition of the religious complex, and continued through the following decades until reconstruction finally won out and took place from the 1950s to 1970. Following, we analyse and discuss the correlation of forces that led to this reconstruction from three aspects: toponymy courses of action, the circulation of the image of the church and the destroyed college, and political actions.

## THE PERSISTENCE OF JESUIT TERMINOLOGY IN LOCAL TOPONYMY

The attempt to erase the Jesuit memory from the heart of the city of São Paulo goes back to the very process of their expulsion. Three toponyms designated the site: *Colégio* (College), *Igreja do Colégio* (College Church) and *Pátio do Colégio* (College) designating the square in front of the buildings. With the departure of the Jesuits, the set was renamed: *Colégio* officially became *Palácio dos Governadores* (Palace of the Governors)and *Pátio do Colégio* became *Largo do Palácio* (Palace Square). The *Igreja do Colégio* received the religious denomination *Igreja do Bom Jesus* (Church of Good Lord Jesus) and also a civil name, *Capela do Palácio* (Chapel of the Palace) or *Capela Imperial* (Imperial Chapel) (Donato 215). Therefore, Jesuit toponymy was officially already gone in the eighteenth century. But a series of maps and images from the following decades shows the resilience of the traditional names.

In the *Plan of the City of São Paulo*, produced by Portuguese engineer,Rufino Felizardo e Costa, between 1807 and 1810, the city is portrayed as a 'small town that left the eighteenth century rehearsing its first steps towards the intensification of an urban life' (Campos, E. 4). In this map, the toponym *Largo do Palácio* does not appear, and the church was named *Colegio dos Extintos Jesuítas*(Former Jesuit College).

In parallel, more images of the set of buildings were being created, like those produced by Thomas Ender[6] (1818) and Debret[7] (1827), portraying the set as the *Palácio dos Governadores*, mentioning neither the old college nor the church. Cartographer Felizardo, a Portuguese resident in São Paulo who was probably already familiar with the customs of São Paulo society, observed the place with its old name, taking into consideration the Jesuit memory, while the foreign painters Ender and Debret used its official name.

In 1841, engineer Carlos Abrão Bresser printed a map depicting *Pallacio da Presidencia* (President's Palace) and *Igreja do Collegio*. In the same year, another map from Ruffino Felizardo applies both names to the set: *Palácio do Governo* for the former college and *Colegio dos Jesuitas* for the church. The 1842 *Carta da Capital de São Paulo* also depicts the former college as *Palacio* and the church as *Collegio*.

In 1847 Bresser produced another map called *Map of the City of São Paulo and its Outskirts*, attributing *Palácio* as the toponym for the entire set. Concomitantly to the production of the map, Miguelzinho Dutra[8] portrayed the set as *Largo do Palácio* and *Igreja do Colégio dos Jesuítas*.

In 1862, the first photographic records appeared, images by Militão Augusto de Azevedo. In his *Comparative Album of the City of São Paulo*, Militão named the building *Palácio dos Governadores*, and the church is designated as *Church and Convent of the Provincial College serving as the Palace of the Governors, General and Provincial Treasury Department, Provincial Assembly, House of Taxes and Post* (Figure 4.1). The *Planta da Cidade de São Paulo* of 1868 depicts the *Palácio do Governo* (Government Palace) and the *Igreja do Colégio*.

In 1877, the territory is represented by French lithographer Jules Martin and engineer Fernando de Albuquerque in the *Map of the Capital City of the Province of São Paulo*, whose intention was to guide people travelling from inland who came to do business in São Paulo, as well as foreigners who came to work in the city (Campos, E. 11). In this map, the L-shaped square receives two toponyms, *Largo do Palácio* for the larger space in front of the Governor's Palace/Former College and *Páteo do Collegio* for the smaller and lateral side of the square. In this map the church is named *Igreja do Collegio*. This indicates, once again, the appropriation of the toponyms related to Jesuit memory by both São Paulo residents and visitors, which can also be seen in the publication of a letter in *O Estado de São Paulo* newspaper in 1875.[9] This report describes a visit by Emperor Dom Pedro II to the city after attending mass at the *Church of Sé*, while '6–8 horses headed towards *Pateo do Collegio* preceding the carriage of the Emperor, they stopped, entered the church, and went on with their prayers as the good Christians they are' (*Carta de Uma Roceira* 09).

The 1881 *Plan of the City of São Paulo*, organized by the Water and Sewage Company, indicates *Palácio do Governo* and *Igreja do Collegio* perpendicular to *Beco do Collegio* (College Lane). Both names, *Beco do Collegio* and *Pateo do Collegio,* can be found in advertisements published in the press in the late nineteenth century.[10]

After the establishment of the Republic in 1889, references to the Jesuit college and church were no longer seen on city maps. The *Plan of the Capital of the State of S. Paulo*, from 1890 by Jules Martin, presents a line dividing the palace and the church, but does not name the Church, although it still existed at that time, in a sign of the process of erasure that was going on at the period. A map of 1897 by engineer Gomes Cardim shows the whole building named *Palacio* and no sign of the church that had been demolished in the previous year. The Jesuit toponym remained dormant for about four decades.

On 13 November 1930, *Largo do Palácio* was again renamed with toponym *Praça João Pessoa*, as stated in the *Topographic Map of the Municipality of São Paulo* from 1930.[11] The name was given in honour of the politician and country's vice presidential candidate, murdered that same year. João Pessoa was a politician linked to Getúlio Vargas, the President who later became a nationalist dictator whose political career was erected in opposition to the São Paulo elite. The new name assigned to *Largo do Palácio* therefore represented the national central government seeking to impose itself on the symbolic centre of the capital of São Paulo, the province that most opposed the centralization of political power.

The tension between the federal government and the São Paulo elite culminated in a true civil war in 1932, in which the *Paulistas* (the name given to those born in the province of São Paulo) led an armed uprising against what they considered to be institutional abuse on the part of the federal power. The revolt was brutally repressed, resulting in a resounding defeat of the São Paulo insurgents and a death toll in the state of about 1,000. Since then, the episode has been named by the *Paulistas* as the Constitutional Revolution, since they were fighting for the declaration of a new constitution.[12]

In 1934, the federal government effectively decreed a new constitution, and in the regional narrative, their defeat was turned into a victory in the fight for their rights; the victims of the revolt were turned into martyrs. Since then in São Paulo, an intense cult of the memory of the 1932 revolt has formed, as well as for local specificities of the state considered to be a supposed defender of the country's freedom – a cult that could be considered to be a kind of 'sub-nationalism'.

The conflict of 1932 also led to important changes in the local toponymy. Among other examples, *Avenida Anhangabaú*, a model avenue from the 1930s, was renamed *Avenida 9 de Julho*, commemorating the day the revolt began.

In this framework, *Praça João Pessoa* was also the object of concern due to its name's relation to the central power. In 1934, the site was the stage of an episode that became known as the *Battle of the Signs*, characterized by the late-night replacement of the *Praça João Pessoa* sign with another one bearing the old Jesuit name *Patio do Colégio,* an action repeated every time the municipality replaced the square's sign with one bearing its official name (Donato 235). The choice by the locals of the Jesuit term *Patio do Colégio* as a battle flag – and not its secular name, *Largo do Palácio* – reveals the rebirth of a search for traditional elements, specific to local story, that would help produce a symbolic counterpart to that currently being defended: the exceptional nature of the São Paulo saga, taming the interior of Brazil. Along with the *Bandeirante* (explorers), the Jesuits were recovered in the regional narratives as mythical figures, privileged in the reconstruction of São Paulo's heroic past (Ferreira, A. C. 111).

The dispute over the square's name was resolved in favour of the insubordinates. In 1936, a municipal act made *Pateo do Collegio* the site's official name: both terms *Praça João Pessoa* and *Largo do Palácio* were defeated by the traditional terminology, which continues to this day. The resolution reveals the relevance of the name to a set of social organizations and local institutions:

> *Considering that this city was founded [ … ], around the Church of the College of the Good Lord Jesus; considering that this square had had different names, one of the oldest being Pateo do Collegio, given spontaneously by the first settlers; considering that to bring back that location's original and symbolic name is to meet a collective aspiration, also meeting the request of more than one association, including our Historic and Geographical Institute and the Paulista Academy of Letters; considering that this denomination is equivalent to a homage to the undefeated builders of this land, and that it evokes tradition and history, thus it comes from the heroic time of conquest, of catechesis, and of the founding of the city of São Paulo, [the Governor] Decrees [ … ] the square that is home to the Palace of the Governors will be renamed Pateo do Collegio. (Pateo do Collegio 8)*

Thus, the first movement for the reconstruction of the Jesuit College and Church came about through protracted resistance to change the secular nomenclature throughout the nineteenth century, followed by a more targeted act of political revolt in the 1930s, which resulted in a return of the site's traditional name.

The 1936 return of *Páteo do Collegio* toponym as the site's official nomenclature was both the result of a long process and the first administrative act of its recovery, a fact that would culminate in the physical reconstruction of the set a few decades later. To the toponymy battle, other dimensions of the struggle for the reconstruction of the Jesuit set were added in the following decades.

## THE CIRCULATION OF IMAGES OF THE DESTROYED CHURCH AND COLLEGE

After the expulsion of the Jesuits in the mid-eighteenth century, the features of the set of buildings remained unchanged for over a century. In the final decades of the nineteenth century, modernizing renovations were made that gave the college building a neo-classical façade and showed the transition of the city's monumental buildings from a Portuguese style towards a more French and internationalized imagery. These changes were made only on the palace building, thus preserving the church's original colonial façade. A nostalgia-inspired backlash did not happen until the end of the nineteenth century.

But the destruction of the church in 1896 did not go unscathed in local lore. Some time after its destruction, images of the destroyed colonial buildings began to circulate, passed around as a conservative and nostalgic reaction over the loss of traditional values. Nostalgia around the destroyed church also brought to mind the image of the college's colonial façade, which had been replaced by the Palace of the Governors.

The images of the destroyed buildings began to spread in the 1910s. At that time, this was due largely to parts of the São Paulo intellectual elite, centered on two institutions founded in the late nineteenth century and dedicated to the regional history of São Paulo: the *Museu Paulista* (Paulista Museum) and the Historical and Geographical Institute of São Paulo. The two institutions operated in their realms of expertise – the Institute with a greater focus on written sources, and the Museum emphasizing material culture and museology – both, nonetheless, focusing on the contributions *Paulistas* made to Brazil's history. The movement was more than a mere intellectual exercise: in the early twentieth century, the São Paulo elite was the most powerful economic group in the country and led the production of coffee, Brazil's largest export. In the beginning of the twentieth century, over half of the coffee consumed in the world was produced in São Paulo and the local elite needed symbolic references to support its economic prominence.

The Museum, the Institute and other social groups led this movement from two main approaches: first of all, through an appreciation of the *Bandeirante*, a traditional mythical figure representing as a homogeneous group the *Paulistas* who explored the interior of Brazil in a quest to capture indigenous and mineral wealth between the sixteenth and eighteenth centuries. Secondly, they praised images of colonial São Paulo, the poor but brave village from which the *Bandeirantes* set off and to where they returned. This made the local historical narratives different from those promoted by other institutions, such as the National History Museum in Rio de Janeiro, which used as a key milestone the independence from Portugal rather than the exploration of territory.

The Jesuits still did not appear as the protagonists in this story at that time, partly because of their history of friction with the *Paulistas* during all the colonial period, mostly concerning theirs resistance against Indian slavery. Nonetheless, the destroyed church and college, as they were famed landmarks in the city, appear as a highlight in a series of representations from the 1910s and 20s aimed at glorifying São Paulo's colonial past, such as an oil painting by Benedito Calixto, currently in the collection of the Sacred Art Museum of São Paulo; a painting by Wasth Rodrigues commissioned by Alfonso E. Taunay, the Director of the Paulista Museum from a drawing by Thomas Ender from 1819; and a model of the city of São Paulo in 1840, on display in a Paulista Museum room built for the exhibition commemorating the centennial jubilee of Brazilian Independence since 1922 (Brefe 87–96).

The images and institutional atmosphere that promoted the special features of São Paulo and the *Paulistas* in the historical construction of the country took on new meaning in the 1930s. After the success of the 1917 socialist revolution in the Soviet Union, the threat of new revolutions crossed the Atlantic and the mobilization of socialist, Communist and anarchist movements were expressed in particularly intense ways in São Paulo, the home of industrialization and the largest and most active working class in the country. Catholic values began to be evoked by a group of conservative intellectuals who attacked the secular nature of Marxist proposals, presenting them as an alternative to the growing Marxist revolutionary trend. Support for Catholic values increased even more after the rise of the Nazi and fascist regimes in Europe in the second half of the 1930s, which were also seen as a secular threat by Brazilian conservatives (Rodeghero 478–80).

While the so-called 'First Republic'[13] had separated Church and State in a radical way, the nationalist government that ascended to power in the 1930s promoted certain conciliation between the two realms, allowing, for example, the return of religious education to public schools.[14] Since the 1920s (reverberations of the Soviet Revolution can be seen here), great religious landmarks returned to the urban setting with the proposition to erect Christ the Redeemer on Corcovado hill in Rio de Janeiro, which was inaugurated in 1931 (Grinberg 57–72). Thus, the official glorification of religious elements was once again possible, and the return of the toponym *Pateo do Collegio* should also be understood in this context.

With the conciliation between the official discourse and religion, the memory of the Society of Jesus was also redeemed in Brazil. In 1937, the same year the national historical heritage service was created, architect Lucio Costa was sent to the region of the former Jesuit missions in southern Brazil,[15] and in 1938 the ruins of the missions were registered as a heritage site, among the first registrations in the country. In 1940 the Missions Museum was inaugurated in the mission of San Miguel in Rio de Grande do Sul province, designed by the same architect. That can be considered as an example of the sealing of peace between Brazilian modernists and the country's Jesuit past. In place of obfuscation, the Jesuits went down in history as promoters of civilizing values.

The fact that the city of São Paulo was founded initially as a college inspired a series of representations praising the local vocation for education. One of the largest institutional achievements of the 1930s in the province, the University of São Paulo, adopted the image of the college ex-libris: the college is in the backdrop, on the tree's trunk is the inscription 'Colégio, 1554', and in its leafy crown the words 'Universidade, 1934' (Figure 4.2). That same year, the Law School was renovated after a fire, and the image of the college was reproduced on one of its stained glass windows. In 1939, the Marina Cintra public school was inaugurated on *Rua da Consolação*, one of the city's main avenues, and its modernist façade was decorated with a panel of tiles showing one Jesuit catechizing indigenous children, the colonial college in the background.

Thus, by the beginning of the Second World War, not just the square had been renamed after the city's Jesuit memory, but also images of the destroyed church and college were circulating broadly, carrying different and complementary messages:

4.2 Ex-libris of the Philosophy School of the University of São Paulo in 1934, linking the grandeur of the University with the educational origins of the city of São Paulo, represented by the then-destroyed Jesuit College
*Source:* Ex-Libris, Universidade de São Paulo, 1934. José Rodrigues.

São Paulo's heroic past legitimating its present economic dominance; anti-Communism; anti-Nazi-Fascism; the city's scholarly and scientific vocation. All these dimensions set foundations for the physical reconstruction that was to begin in the following years.

## POLITICAL ACTION AND FUNDRAISING

In the early 1940s, the nostalgic quotations of the Jesuit and colonial shape of *Largo do Palacio* turned into an actual movement of advocacy for the effective reconstruction of the religious complex.

A sort of symbolic crisis had already begun a few decades before for the buildings. After losing their religious significance, the buildings would also lose their meaning as a symbol of political power. In 1908, the governor's residence moved to *Palácio dos Campos Elíseos*, a grander building in a residential neighbourhood favoured by the city's elite from the western zone, which would later consolidate itself as the city's elegant area and remain such throughout the twentieth century. In 1930 – the same year *Largo do Palácio* was renamed *Praça João Pessoa* – all governmental functions were transferred to Campos Elíseos. The Department of Education remained in the Palace, a bureaucratic office with much less symbolic stature than the governor's palace (Ferraz de Lima 68–9).

Five years after the return of the toponym *Pateo do Collegio*, in 1941, José Mariano Filho, an intellectual from Rio de Janeiro who since the 1920s had advocated the return of Brazilian architectural imagery to a colonial architectural lexicon, publicly defended the reconstruction of the Jesuit church in São Paulo, a sentiment that was already circulating in the media at that time, with Mariano noting 'the enthusiasm with which it was being welcomed by all intellectuals throughout the city of São Paulo'.

> In 13 years, São Paulo will celebrate the 400 year anniversary of its founding, and an act of the highest cultural significance would be the full reconstitution of the Jesuit church, and a wing of the old college, on the exact site where it originally existed, so that the first mass could once again be prayed in the Piratininga plateau in the small oratory made 'of sticks and clay' built by the indigenous people to Joseph de Anchieta. São Paulo has a sacred debt to pay to the Society of Jesus. And the time has come. (Mariano apud Cardim Filho 1975  90)

In the same year, an official pro-church committee was created that consisted of important figures from the city, such as former mayor Luiz Inácio de Anhaia Melo. Nevertheless, it did not achieve any practical results (Donato 2008 242).

The Catholics and conservative intellectuals found the Association of Former Jesuit Students (ASIA) proved to be a very effective political tool in advocating for the reconstruction of the compound. Organized in different parts of the world, they held regular meetings together with the Jesuits, whose goal was to continue their Christian and Ignatian training, thus perpetuating the memory of the Society of Jesus. ASIA São Paulo was founded on 17 November 1926. The group was composed of notable alumni from the Colégio São Luis and actively participated in movements to spread the Jesuit memory in São Paulo, uniting some of the most influential members of São Paulo society.[16]

The dispute over the reconstruction of the Church and College was quiet for a few years, while the Catholic agenda gained public prominence in the city and country. In 1942, the 4th National Eucharistic Congress was held in São Paulo and crowds gathered in the streets in support of the Catholic agenda. The agenda for the reconstruction of the College and Church was not mentioned during the preparation of the Congress, possibly because the city was still fully campaigning for the construction of its immense cathedral, which had been under construction since 1911 (*São Paulo no IV congresso Eucarístico Nacional*).

In 1948, the discussion over the reconstruction of the Jesuit buildings appeared for the first time in the minutes of the Alumni meeting, where a committee was formed to take the matter forward.[17] In the 1950s, with the celebrations of the city's 400-year anniversary, discussions intensified. There was a consensus that the college area needed changes in order to evoke the birth of the city. The project that seemed to guide public debate was proposed by Carlos Alberto Gomes Cardim Filho, then Head of the São Paulo Urban Planning Department, in which the palace would be demolished and replaced with a skyscraper that would house the City Hall. In this proposal, the church (but not the college) would be rebuilt on a corner of the courtyard, smaller and in a different location than where it had historically existed (Gomes Cardim Filho 180).

In 1952, ASIA's Board of Directors joined the city's 400-year anniversary celebrations, upholding the agenda for the reconstruction of the old church and college. The creation of two committees for the anniversary of the Foundation of the College São Paulo of Piratininga was approved, one holding an executive function, and the other serving as an Honorary Committee.[18] The following year, José Nunes Vilhena, historian and former Jesuit student, made various allegiances, in both politics and the media, to the old college reconstruction movement, calling for a more specific committee within the preparations for the 400-year anniversary celebrations, devoted to the reconstruction of the church and college.[19]

In November 1953, the São Paulo state government decided to completely demolish the Department of Education, and did so diligently (Canado Junior 37). The government knew better what should be destroyed than would be built instead, and two alternatives were posed: the reconstruction of the Jesuit compound or the construction of a more monumental civic centre. Promoters of the civic centre cited the separation between church and state as an argument, and reminded the city that the catholic bishopry had been duly compensated when the church was destroyed in the late nineteenth century.

During the process of demolishing the old Palace of the Governors, ASIA organized a Committee for the protection of the physical remnants of the college, treated by them as relics, and promoted a three-day inspection to ensure the preservation of relevant objects.

José Nunes Vilhena, a former student and the person responsible for monitoring the demolition works, described the process of identifying the different materials when recovering the stone and *taipa* foundation, finding seventeenth-century bricks, ceiling and floor timbers, and wooden and iron nails – and amidst it all, they found a *taipa de pilão* wall, reinforced by rafters, probably dating back to the seventeenth century and which has been preserved.

While an ASIA group sought to preserve what remained of the building, another group of former students used different strategies to defend the reconstruction of the college, such as political alliances with notable alumni in strategic positions in the state of São Paulo. On 21 January 1954, in a major political victory for the former Jesuit students, the land was given back to the Society of Jesus by the state government. The Jesuits got the mandate to build a church, a school and the so-called *Casa de Anchieta*, which was to become a museum. They also took over the property rights to the existing remains and objects on the grounds.[20]

Shortly after the demolition of the palace and the restitution of the land, the group of former students began construction on a full-size replica of the primitive Jesuit Church that marked the beginning of the European and Christian occupation in the region. It was completed in March 1954, made of wood and straw (Salgado 1976 138).

When the celebrations for the 400-year anniversary were finished, in January 1955, the primitive church was dismantled and the group of former students began raising funds to rebuild the college. A movement known as the campaign of *Gratitude to the Founders of São Paulo* was organized, and its main representatives aimed to win over the support of politicians, intellectuals and distinguished members of São Paulo society and sought complete reconstruction of the old Jesuit compound. Using the most strategic media of the time (radio and newspapers), the campaign promoted cocktail parties and dinners, seeking the allegiance of the *Paulistanos of goodwill*,[21] whose patriotic values could support the cause and raise the money necessary for the reconstruction of the Jesuit buildings. The group's main speaker was Altino Arantes, former President of the state of São Paulo, who invited the most distinguished political and social celebrities, thus raising not only monetary donations, but also participants for commissions and work fronts.

From a technical point of view, the reconstruction process was again entrusted to the engineer Carlos Alberto Gomes Cardim Filho, who had recently retired from public service but reformulated the previous project and designed the College and the Church on the exact site where they had originally existed in colonial times. Cardim sought to recover the urban image of the compound based on existing iconography, such as the image by Thomas Ender from the beginning of the nineteenth century and the photography of Militão de Azevedo from the early 1860s. On the other hand, from the point of view of construction techniques and interior design, the modernizing project utilized reinforced concrete and an architectural plan that was different from the original.

Reconstruction was done in two phases, and took two decades (1954–79). The land granted to the Jesuits corresponded to the college and the old Palace of Governors turret, which used to be the Church tower (Canado Junior 56). The grounds where the church nave sat had become the side square of the palace in the early twentieth century, thus creating a particular property problem. According to the Brazilian legal system, it is much more difficult for the public authorities to donate a free area, such as a street or a square, than a building. In the 1960s, Governor Ademar de Barros objected to transferring part of the square's land to the Society of Jesus because of this specific problem. But those advocating the reconstruction of the church claimed that the struggle was not about the privatization of public property, but rather the restitution of land unjustly taken from its previous owners. In parallel, there was also uncertainty about the ownership of the square and whether it belonged to the province or to the municipality, which was resolved in favour of the municipality (Canado Junior 146).

From the point of view of the reconstruction itself, the first public opposition to the reconstruction of the full compound dates back to 1965 from State Representative João Hornos Filho, who stated that 'from a historical point of view, it would be a real crime to add something modern, apocryphal, to what has been there for so many centuries' (Canado Junior 142).

That same year, the campaign for the church's reconstruction received support from a spiritual order: a femur of Jesuit José de Anchieta, considered the Apostle of Brazil and whose beatification process was underway, was transferred to São Paulo. The femur was treated as a sacred relic, received with all honours and positioned in a chapel constructed under the tower, the only part of the church that had already been built, thus triggering the reintegration of the site into the Catholic mystical temporality and contributing to its re-consecration.[22]

The uncertainty regarding the land where the church was to be built lasted until 1968, when Mayor Paulo Maluf – also a former pupil of the Jesuits – ceded the land to the Society of Jesus for the reconstruction.[23]

While in the 1950s the reconstruction proposal had not awakened explicit opposition from the historical heritage specialists, by the 1970s the situation had changed drastically. Heritage experts had the corresponding provisions of the Venice Charter of 1964, and some of them spoke in the media and in public opinion forums denouncing what they saw as a grotesque pastiche that disregarded the material evidence still at the site.

In July of 1975, with the college already built and the church under construction, the Society of Jesus received a notification from Condephaat (the provincial agency for Heritage, created in 1968) requesting suspension of the construction underway due to a plan to register the location as an archaeological site. The arguments condemned any attempts to rebuild the totally destroyed buildings, and claimed that there was not enough documentation to faithfully reconstruct the old compound.[24]

Defending the college church 'not only as a sanctuary of the first generations of settlers, but also as a landmark of the foundation of the city', Cesar Salgado wrote the document *In Defense of the Historical Heritage of the College Courtyard*, which sought to respond to the arguments raised by Condephaat, comparing the construction of the Church of the Good Lord Jesus to other cases he considered similar for 'being rebuilt in the image and likeness of the original buildings, and on the same foundations' (Salgado 253–73); among them, the former student cited other reconstruction processes: the Goethe House in Frankfurt, the medieval churches of London, the Church of St Nicholas of Sangemini, the Campanile of Venice and the Golden Pavilion in Kyoto.

4.3   Jesuit
compound
rebuilt between
1954 and 1979
*Source:* João Carlos
Santos Kuhn.

On 9 March 1976, Provincial Governor Paulo Egídio Martins authorized the construction of the church within the previously established limits, ending discussions with Condephaat and other social actors that opposed the donation for the reconstruction of the church.

A new fundraising campaign was begun, but which proved to be quite troublesome and was only resolved between 1976 and 1979 when the city donated a significant portion of the necessary funds for the completion of the building (Canado Junior 199).

## FINAL REMARKS

The case of the Jesuit college and church in São Paulo poses a quite radical example of reconstruction. Beyond traditional architectural and spatial forms, here what was at stake was an amazing and quite unique process of devolution to the Jesuits of a piece of land that had been taken from them almost two centuries before, in a process of de-secularization and re-sacralization of a space that took place in the very heart of the city and its symbolic centre. The common linearity in the transition of spaces in Brazilian cities in the nineteenth and twentieth centuries, 'from the sacred to the profane' was not observed (Marx 7–14). In this case the movement was more intricate, going from the sacred to the profane, then again to the sacred.

The reconstruction of the Jesuit buildings in the heart of São Paulo is often negatively viewed by scholars and intellectuals, who use of terms like 'pastiche', 'simulacrum' and 'dramatization'. In fact, if we go by the contemporary schools of conservation and restoration, then the reconstruction of the buildings is not a good example of how to properly treat material evidence of the past.

On the other hand, if we consider the reconstruction as the outcome of a long process in which collective representations and different social groups worked to interpret historical events and be a part of them, the scenario presents itself as something much more complex.

The technical-political articulation that culminated in the overwhelming lobbying for the reconstruction of the Jesuit buildings between the 1950s and 70s was highly intricate and effective. It was founded over an alliance of decision makers with deeply-rooted beliefs and represented almost all segments of society and with sophisticated political, legal and financial strategies.

If we consider all the dimensions of the resistance of the Jesuit memory in the college courtyard since the expulsion of priests, we will come to an image of an almost awesome power. The church reconstruction project successively defeated Marques do Pombal, the secular liberals of the first republic, the dictator Getulio Vargas, the chief architect of the first republic, Ramos de Azevedo, the Plan of Avenues by Prestes Maia, the populist governor Ademar de Barros, the technicians of the historical heritage bearing the Venice Charter – and, of course, a lack of resources that is structural of this type of campaign in Brazil.

Opponents of reconstruction, on the other hand, presented themselves in the different debates with a notable weakness: those who defended the permanence of the site as a public and secular site were defeated in the 1950s, and those who advocated a more rigourous approach from the point of view of restoration techniques and respect for the existing built matter were defeated in the 1970s. They were late stating their position, as they had no defined social base and little ability to propose choices that could compete with the reconstruction, from a symbolic point of view. They were not able to build productive dialogues with conservative segments of society, nor did they know how to relate to a popular imagery which increasingly became very favourable to the reconstruction.

## BIBLIOGRAPHY

Brefe, Ana Claudia Fonseca. *O Museu Paulista: Affonso De Taunay E a Memória Nacional 1917–1945*. São Paulo, SP: Museu Paulista, Universidade de São Paulo, 2005. Print.

Campos, Eudes. *São Paulo Antigo: Plantas da Cidade*. Informativo Arquivo Histórico Municipal, 2008. Web. 30 Nov. 2014 <http://www.arquivohistorico.sp.gov.br /info/ info20/index.html>.

Campos, Cândido Malta. *Os Rumos da Cidade– Urbanismo e Modernização em São Paulo*. São Paulo: SENAC, 2002. Print.

Canado Junior, Roberto dos Santos. *Embates Pela Memória: a Reconstrução do Conjunto Jesuítico do Pátio do Colégio (1941–1979)*. São Paulo: Dissertação de Mestrado, FAU-USP, 2014. Print.

Candido, Antônio. *A Educação Pela Noite e Outros Ensaios*. São Paulo: Ática, 1989. Print.

Cardim Filho, Carlos Alberto Gomes. 'Pátio do Colégio'. *Revista do Arquivo Municipal* 187(1975): 83–97. Print.

Cardim Filho, Gomes. 'O Pateo do Colegioe o Centenário'. *Revista Acropole* 151(1950): 180–81. Print.

Cymbalista, Renato. *Cidade dos Vivos: Arquitetura e Atitudes Perante a Mortenos Cemitérios Paulistas.* São Paulo: Annablume, 2002. Print.

Donato, Hernani. *Pateo do Collegio: Coração de São Paulo.* São Paulo: Loyola, 2008. Print.

Ferraz de Lima, Solange. 'Pátio do Colégio, Largo do Palácio'. *Anais do Museu Paulista* 6–7.1 (1999): 61–81. Print.

Ferreira, Antonio Celso. *A Epopeia Bandeirante: Letrados, Instituições, Invenção Histórica (1870–1940).* São Paulo: Editoria Unesp, 2001. Print.

Ferreira, Roberto Martins. *Organização e Poder: Análise do Discurso Anticomunista do Exército Brasileiro.* São Paulo: Annablume, 2005. Print.

Grinberg, Lúcia. 'República Católica: Cristo Redentor.' *Cidade Vaidos: Imagens Urbanas do Rio de Janeiro.* Ed. Paulo Knauss. Rio de Janeiro: Sette Letras, 1999. 57–72. Print.

Lopreato, Christina Roquette. *O Espírito da Revolta: a Greve Geral Anarquista de 1917.* São Paulo: Annablume, 2000. Print.

Marx, Murillo. *Nosso Chão, do Sagrado ao Profano.* São Paulo: Editora da Universidade de São Paulo, 1989. Print.

Miceli, Sergio. *Intelectuais à Brasileira.* São Paulo: Companhia das Letras, 2001. Print.

Rodeghero, Carla Simone. 'Religião e Patriotismo: o Anticomunismo Católico nos Estados Unidos e no Brasil nos Anos da Guerra Fria'. *Revista Brasileira de História* 22. 44 (2002): 463–88. Print.

Rolnik, Raquel. *A Cidade e a Lei.* São Paulo: Studio Nobel/FAPESP, 1997. Print.

Salgado, César. *O PátiodoColégio: História de uma Igreja e de uma Escola.* São Paulo: Gráfica Municipal de São Paulo, 1976. Print.

*São Paulo no IV Congresso Eucarístico Nacional (1942).* São Paulo: Oficina Gráfica da Ave Maria, 1942. Print.

## Periodicals

*Carta de Uma Roceira.* O Estado de São Paulo. 29 Aug. 1975. p. 9.

*Pateo do Collegio.* O Estado de São Paulo. 2 Apr. 1936. p. 8.

## Cartography

*São Paulo Antigo: Plantas da Cidade* (1954). São Paulo: Comissão do IV Centenário/ Editora Melhoramentos.

## NOTES

1　*Craveiro* is a Portuguese linear unit of measurement; 1 craveiro equals 5 feet, or 1.52 metres. 10 × 14 *craveiros* equals 15.2 × 21.2 metres.

2　*Taipa de Pilão.*Luso-Brazilian version of traditional adobe construction painted with lime. Also rammed earth.

3    In the second half of nineteenth century, São Paulo experienced profound transformations with the growth of coffee plantations in the hinterlands of the province and the construction of the country's most comprehensive railway system which led to the capital city, then on to the port of Santos.

4    *Terreiro* or *adro* (yard) corresponds to a square in front of religious buildings in Portuguese-origin cities.

5    Brazil had the only monarchic experience in the Americas; it was ruled by two Emperors from 1822 to 1889, the year the Republic was established.

6    Thomas Ender was sent to Brazil by Princess Leopoldina of Hapsburg, who came to Brazil to marry Prince Pedro de Alcântara in the early nineteenth century, and ordered a scientific natural history expedition in order to gather information about the country and develop a Brazilian museum in Vienna.

7    Jean Baptiste Debret was born in Paris, France on 18 April 1768. A graduate of the Academy of Fine Arts in Paris, Debret was a member of the French Artistic Mission in Brazil, organized at the request of King Dom João VI and led by Joachim Lebreton.

8    Miguel Arcanjo Benício de Assumpção Dutra (1812–75). Painter, sculptor, goldsmith, architect, poet, engraver, church decorator and musician. In 1841, Miguelzinho completed the drawings that border the first map of the city of São Paulo, one of the most important sources in São Paulo iconographic documentation of the nineteenth century.

9    *O Estado de São Paulo*, 26 August 1875.

10   *O Estado de São Paulo*, 26 August 1886.

11   Printed cartographic document produced in 1930 by the company *Sara Brasil*.

12   The 1932 crisis was one of the most important episodes in the history and mind of São Paulo. Although the watchwords were political and institutional, there was a struggle for the province's economic independence behind them, the main producer and exporter of coffee, and the richest region in the country.

13   'First Republic' or 'Old Republic' designates the period of Republican liberal government in Brazil from 1889 to 1930.

14   Decree no. 19.941, 30 April 1931.

15   The Jesuit missions in America were settlements organized by the Society of Jesus among Guarani Indians in the seventeenth and eighteenth centuries, established for the purposes of evangelization and civilization. The Jesuit missions were destroyed in the mid-eighteenth century when the Portuguese and Spanish empires fought to banish the Jesuits from the entire continent.

16   Association of Former Jesuit Students. São Goncalo Church. Minutes of the meeting held on 17 November 1926. p. 5.

17   Association of Former Jesuit Students. Colégio São Luis. Minutes of the meeting held on 28 November 1948. p. 30.

18   Board of Directors of the Association of Former Jesuit Students. Colégio São Luis. Minutes of the meeting held on 31 May 1953.

19   Association of Former Jesuit Students. Colégio São Luis. Minutes of the meeting held on 21 June 1953.

20   'The donation of the land is accompanied by the relics therein, but the grantee is responsible for conservation, in a proper place, and for building a new college in

São Paulo, and a church annex to it, as much as possible within the limits of the initial foundations, and for reproducing in a perfect replica, the initial act in the foundation of the city of São Paulo, executing the laying of the cornerstone for the work that will perpetuate the dearest tradition of the people of São Paulo on the occasion of its 400-year anniversary, to be celebrated on 25 January 1954' (State Law 2.568 / 1954, art. 2).

21    The term *Paulistanos of goodwill* was often used by Altino Atantes in his speeches seeking new supporters for the Gratitude to the Founders of São Paulo Campaign.

22    *O Estado de Sao Paulo* 24 March 1966. p. 15.

23    Municipal Law 7356/1969.

24    Condephaat – Official Letter SE-52/75.

# 5

# The Issue of 'Identical Reconstruction' on French Heritage Sites: Architectural Cloning, Alternate History and Tourism

*Julien Bastoen*

## INTRODUCTION

In the introduction of a work published in 2007, Spanish art historian Ascención Hernández Martínez wrote, 'If we observe the recent proliferation of architectural clones, from old buildings or key buildings of the architecture of the twentieth century, we must conclude that this is not a fragmentary or episodical phenomenon, but a true trend of which the motivations should be known' (14).[1] The ICOMOS adopted a resolution, in its 17th general assembly in Paris in 2011, to initiate a debate on this growing phenomenon, 'noting the increasing disregard of existing theoretical principles for the justification of reconstruction, and a new tendency towards significant commercialization of reconstruction activities' (Resolution 17GA 2011/39). The ICOMOS ultimately conducted an online survey in 2014 on the permissibility and standards for reconstruction of monuments and sites.

France did not escape this phenomenon. Several reconstruction plans of extinct buildings stirred controversy, most of them related to the past French monarchic regimes and situated in highly protected heritage districts, two of them even being World Heritage sites. The recent reconstruction of the royal gate of the Palace of Versailles and the reconstruction plans for both the Tuileries Palace, in Paris, and the Saint-Cloud castle, in the western suburbs of Paris, question the occidental codes of authenticity and even the ICOMOS charters. Such cases are quite different from the recent reconstruction of the old parliament buildings in Rennes and the Lunéville Palace – both significant landmarks damaged by fires in 1994 and 2003, respectively. Indeed, 'when it comes to rebuild from scratch when time has passed – more than half a century or more – deep motivations and methods of implementations are quite different' (Pallot-Frossard 86).[2]

Considering a wide range of reconstruction plans, Dawans and Houbart have shown that a 'departure from the principles and spirit of the main twentieth-century conservation charters seems to encounter the mainstream of cultural capitalism and one of its major economical sectors: tourism. In many cases, this very concrete reason is obviously more important than any identity consideration and reduces the distance between heritage and theme parks' (Dawans). Rather than considering the French cases using concepts developed by analytical philosophy and semiology, we would like to understand who the stakeholders of these projects are and what justification narratives and funding models they use. Is this phenomenon structured? Is there a professional network calling for such reconstruction involving different categories of individual and collective actors? Is it possible to link this phenomenon with a growing nostalgia for monarchy?

## CASE STUDY #1: ROYAL GATE OF PALACE OF VERSAILLES

The royal gate of Versailles was a monumental barrier built between 1679 and 1682 during the second phase of expansion of the palace by architect Jules Hardouin-Mansart. The function of this structure was to separate the so-called *avant-cour* (now called *cour d'honneur*) from the *cour* (now called *cour royale*) to impress visitors and enhance the sacredness of the king. This gate replaced another built by architect Le Vau between 1662 and 1664 under the reign of Sun King Louis XIV. We can see the first version of the gate on a canvas painted in 1668 by Pierre Patel and the second version on a canvas painted by Pierre-Denis Martin in 1722.[3]

In 1771, one of the wings built by Louis Le Vau and transformed by Jules Hardouin-Mansart was already falling apart. King Louis XV asked architect Ange-Jacques Gabriel to rebuild that wing. In order to make that possible, Gabriel dismantled the north part of the royal gate, and replaced it by a wooden palisade and temporary buildings for the workers. The royal gate should have been restored and rebuilt after the new wing was completed. However, that did not happen. After the Revolution, in 1789, some elements of the remaining part of the gate were restored. Finally, the temporary palisade and the remaining part of the royal gate were both removed in 1794 in compliance with an order of deputy Joseph-Augustin Crassous. A few years later, Napoleon I considered implementing the plans that Gabriel couldn't complete before the Revolution; King Louis XVIII finally ordered the reconstruction of the south wing for a symmetrical effect with the north wing rebuilt by Gabriel; the Dufour pavilion was then built in 1818–20. Between 1835 and 1837, King Louis-Philippe ordered that an equestrian statue of King Louis XIV be erected where the royal gate once stood. It was completed in June 1836.

Although former curator Daniel Meyer called for the removal of this equestrian statue and for the rebuilding of the royal gate in 1978, preliminary studies were not carried out before 2002. In 2003, proposals for the reconstruction of the royal gate were included in a €500 million master plan, to be spread over 17 years, which was approved by the French National Heritage Preservation Commission and the French government. The reconstruction of the royal gate was finally carried out between 2006 and 2008, thanks to a €3.5 million sponsorship from Monnoyeur, the largest French retailer of machinery for construction and industry, which celebrated its centenary in 2006 (Rykner 2007). Frédéric Didier, chief architect of the National Heritage, designed the new old gate. He is, along with colleague Jacques Moulin, a co-architect in charge of the gardens and park of Versailles and of the palaces of Trianon, and a partner of the architectural firm 2BDM based in Paris and Versailles. Both architects are known for their uninhibited approach towards restoration and reconstruction. The gate was rebuilt by the construction company 3dPierre, which was involved in the construction of the Château Louis XIV, a replica of Vaux-le-Vicomte, another prominent seventeenth-century palace in the Paris area.[4]

The justification of the recreation of the royal gate at Versailles seems to satisfy mainly logistics and monitoring needs. It has been thought of as the keystone of a new visitor reception system. It will be soon followed by the complete transformation of the interiors of pavilion Dufour in order to create new visitor facilities and enable more efficient control of the increasing visitor flow. The problem is that nobody knows the exact number of visitors.

The project bears a strong resemblance to the cloning technique that John Hammond used to recreate T-Rex and velociraptors for his *Jurassic Park* by extracting the DNA of dinosaurs from mosquitoes that had been preserved in amber. However, as the strands of DNA were incomplete, DNA from frogs was used to fill in the gaps. In the case of Versailles, architect Frédéric Didier, with the help of archaeologists and historians, had to take a two-phase approach. His first task was to identify and locate all the visual sources of the royal gate, such as engravings, drawings and paintings within the French borders or abroad. His second task was to excavate the foundations of the royal gate, and a short excavation campaign was carried out in 2006. Nevertheless, the total lack of convergence between the available materials didn't make it possible to establish a reliable and indisputable representation of the second version of the royal gate as close as possible to Hardouin-Mansart's design. In other words, as the strands of DNA were incomplete, the architect had to fill in the gaps, not with DNA from frogs of course, but drawing on features of comparable seventeenth-century gates situated in the Paris area. The architect had a mixed attitude towards his own historical makeshift job; indeed, in 2007, he gave an assurance that the recreation of the royal gate at Versailles was 'rigorous' and 'supported by a solid and indisputable basis' (Didier 12).[5] One year later, after some architectural historians questioned the reliability of primary sources, Frédéric Didier asserted to historian Franck Ferrand that he had 'spent hours together [along with the blacksmiths] choosing each and every detail, sometimes with the help of archives of course but above all – the archives don't tell us everything – by analogy based on authentic preserved elements either on the Versailles site or on other seventeenth-century gates' (Ferrand).[6]

The Palace of Versailles has been a work in progress since the seventeenth century; it never was and never will be completed despite being transformed and expanded until the twentieth century. For the last three decades, the strategy of both the directors and the architects in charge of the Palace of Versailles has been to make Gabriel's plans come true: they reconfigure Versailles to make it look as close as possible to an idealized eighteenth-century palace. In other words, they don't care about what happened next. In order to legitimize their actions, they use the restoration strategy of a former curator of the palace, Pierre de Nolhac, who was in charge of the collection and the buildings from 1887 to 1920.

5.1    The reconstructed royal gate of the Palace of Versailles
*Source:* Julien Bastoen.

By denying the historical events that shaped the palace through the nineteenth and twentieth centuries, architects (as well as the board of trustees) rewrote history, as if the French Revolution never occurred, at the risk of creating architectural anachronisms. This is evidenced by the fact that the new royal gate starts from a pavilion which was not built yet when the royal gate was definitively removed in 1794. The rebuilding of the royal gate is coupled with the erasing of the nineteenth-century transformations of the court and forecourts, including the relocation of the equestrian statue of King Louis XIV erected on the royal gate site. The paradox is that Pierre de Nolhac, the former curator to whom everyone refers at Versailles, was opposed to any whimsical reconstruction of architectural or landscape elements (De Nolhac). The aim of the current chief architect is to complete what was left incomplete by his predecessors. He considers himself a worthy successor to his famous predecessors to the point that he has even himself represented as a Roman emperor under the inscription SPQR on one of the sculpted groups which were replaced atop the attic on the south wing!

## CASE STUDY #2: TUILERIES PALACE IN PARIS

The Tuileries Palace was built from 1564, that is, from the end of the regency of Queen Catherine de' Medici on a site where tileries were previously established, outside the medieval walls, and about 600 metres west of the Louvre castle. The original plans by Philibert Delorme were continued by Jean Bullant and then construction stopped until the reign of King Henri IV, who decided to build a very long gallery to unite the Tuileries Palace with the Louvre along the north bank of the river Seine. Construction stopped again until King Louis XIV decided to give the palace the symmetry that it lacked. Between the reign of King Louis XIV and the Second Empire, in the middle of the nineteenth century, most of the changes occurred inside the palace. The Tuileries was used as a residence by kings and emperors, including Napoleon I. The last transformations of the palace occurred between 1854 and 1870 during the Second Empire. The Tuileries was, in 1870, a kind of *pot pourri*, that is to say that it was almost impossible to identify the remaining parts of the sixteenth-century palace.

After the fall of the Second Empire in 1870, some events – the siege of Paris by the Prussian army in the winter of 1870–71 and the *Commune*, a short period of insurrection which lasted two months during the spring of 1871 – deeply changed the French political landscape. This insurrection was suppressed with great bloodshed by regime-loyal army units. During the last week of the insurrection, which is now called the 'bloody week', the insurgents set fire to many iconic and public buildings on 23 May 1871, including the City Hall, the ministry of finance and the Tuileries. The palace ruins remained almost untouched for 12 years. Only two wings of the palace were demolished and two pavilions reconstructed in the years 1874–6. Although most of the interiors had been destroyed, the structure shell was still in good condition. However, experts – including Viollet-le-Duc and Léonce Reynaud – selected to conduct a feasibility study on restoration or reconstruction disagreed on the techniques to use for restoration. Yet, the Republican government was trying to prevent any comeback of monarchy rendering the palace – a symbol of monarchy – a problematic building. Ten years after the insurrection, whereas the option of restoration became increasingly dubious because of a degradation of the ruins, members of Parliament were divided into two main camps: the first camp supporting the demolition of the ruins without reconstruction, and the second camp supporting the demolition of the palace and its reconstruction in the spirit of the original sixteenth-century design in order to move the national museum of contemporary art there. Finally, in 1882, both chambers of Parliament approved a law allowing the government to destroy the ruins of the palace (Varnedoe). Paradoxically, the future reconstruction of the palace was not included in the law although the government, though minister Jules Ferry, verbally promised it would be reconstructed as a museum. The ruins were finally demolished in 1883 under the direction of architect Charles Garnier. Meanwhile, the national museum of contemporary art moved to a temporary building, leaving no valid reason for the Tuileries to be rebuilt.

However, many reconstruction plans were considered throughout the last century. Lately, the main supporter of the project had been Alain Boumier, a retired civil engineer who was also the president of the Académie du Second Empire, a historical society whose mission is to rehabilitate the reign of Emperor Napoléon III. Boumier had been calling for building a museum of history of the Louvre and the Tuileries since the 1970s. Boumier was convinced that the best way to pay tribute to Napoleon III would be to rebuild the Tuileries which the emperor had modelled to his own taste. Therefore, he founded the Comité National pour la Reconstruction des Tuileries in 2002 in order to fulfil the 'non-verbalized expectations of the French people'.[7] The Comité National not only lobbied politicians, mostly from right-wing parties, but also tried to shape public opinion through communication of influence, such as using a website, a Facebook group, television, the French and foreign press and organizing conferences and exhibitions. It got the support of intellectuals and scholars, senior officials, diplomats and heritage architects. Among the keynote speakers of the conferences were the two heritage architects then in charge of the castle and gardens of Versailles, Frédéric Didier and Pierre-André Lablaude, as well as Wilhelm von Boddien, the main supporter of the reconstruction of the Berlin City Palace.[8]

In 2006, Renaud Donnedieu de Vabres, the French minister for culture, seemed to pave the way for the reconstruction of the Tuileries. A special commission was set up, composed of only supporters of the project, tasked with assessing its relevance and feasibility. Among the members was novelist Erik Orsenna,[9] who was also a member of the *Académie Française* and the founding president of the Hermione-La Fayette Association, whose aim was to rebuild the Hermione frigate – the ship that allowed La Fayette to join American insurgents in the independence struggle in 1780. Another prominent member of the commission was banker Antoine Bernheim (1924–2012), a kingmaker said to have helped French tycoons Bernard Arnault, Vincent Bolloré and François Pinault to build their business empires.

5.2   The site of the Tuileries Palace, Paris
*Source:* Julien Bastoen.

The key arguments used by the supporters of the reconstruction of the Tuileries Palace are both architectural and urbanistic. They consider that the west part of Paris, along the *Axe Historique* (historical axis), is no longer understandable without the presence of the Tuileries Palace. It is a monumental axis running since the seventeenth century from the central pavilion of the palace to the west. It now includes the Tuileries garden, the Place de la Concorde, the Champs-Elysées, the Arc de Triomphe and the Grande Arche at La Défense. They also say that the absence of the Tuileries Palace makes more obvious the fact that the two wings of the Louvre are not parallel (Boumier). In fact, they re-use arguments from Georges-Eugène Haussmann, who was chosen by Emperor Napoleon III to carry out a massive programme of new boulevards, parks and public works in Paris between 1853 and 1870, and was later involved in a controversy raised by the demolition plans for the Tuileries. However, the reconstruction plans of the Tuileries include the recreation of its forecourt, the so-called Cour du Carrousel presently occupied by gardens. These gardens are a kind of a blind spot and difficult to monitor. Supporters of the project argue that the gardens are fertile ground to many vices such as kidnapping, drug trafficking and even prostitution.[10] The creation of a new segregated space would make it possible to remove the alleged insecurity. It would also be useful to organize events like big historical shows such as those already hosted by Versailles, the Luxembourg gardens in Paris or in the Parc de Sceaux.

The recreation would finally give extra space for the collections of the Louvre and the Museum of Decorative Arts, new reception and meeting areas that could be rented for private events, while a panoramic terrace would be created on top of the dome of the central pavilion.

As far as the Tuileries project is concerned, both the quantity and the convergence of primary sources could make it possible to minimize the part of subjective interpretation. The recreation of the palace could, thus, be possible on a more historically correct basis although an exact copy is definitely impossible to consider. However, this is not what the supporters of the project wanted. The reconstruction has nothing to do with experimental archaeology, but rather with a film setting. If we look closely at the design made by Stéphane Millet, an architect, engineer and designer as well as chairman of architectural firm Clé Millet International, mainly acclaimed for his restoration of historic theatres, the Tuileries would be reconstructed with stone-clad, reinforced, concrete walls and the reconstitution of the interiors would be restricted to a few rooms, staircases and halls (for instance, in museum spaces or reception spaces). The reconstruction, estimated to cost €347 million,[11] would be self-funded by the entrance fee paid by visitors, adopting the funding model of Guédelon in Burgundy, where Michel Guyot, owner of several castles in France, implemented the construction of a medieval-like castle designed by Jacques Moulin in 1997. Guédelon was thought of as a kind of educational theme park using experimental archaeology.

The aim of the reconstruction plans of the Tuileries Palace is not to complete something left incomplete or to build an ideal palace, but to restore the integrity of a once-completed monumental complex. Indeed, before the events of 1871, Emperor Napoleon III had connected the Tuileries Palace with the Louvre on the north side, finally achieving the *Grand Dessein* originally imagined by King Henry IV in the sixteenth century. The demolition of the Tuileries Palace is, thus, seen as a trauma – a gaping wound in the most prestigious area of Paris. Its reconstruction would mean the denial of the insurrection of 1871 as well as of the events that followed. However, as in the case of Versailles, it would also create an architectural anachronism: the new palace would be built between two pavilions that were restored and transformed after the insurrection. Both pavilions are now registered as historic monuments and, therefore, cannot be modified.

Other factors have contributed to bury the project: the hostile attitude of Christine Albanel, the new minister for culture; Bertrand Delanoë, mayor of Paris; Henri Loyrette, chairman of the Louvre, and finally Alain Boumier's death in 2009. Although the project has now lost credibility, the website and Facebook group of the association are still updated by Michel Carmona and Alain Blondy, two former professors at Sorbonne University.

## CASE STUDY #3: SAINT-CLOUD CASTLE

The Saint-Cloud castle dates back to the sixteenth century when the Gondis, a Florentine family of bankers, decided to settle in France and build their home on top of a hill towering over the Seine at a short distance away from the village of Saint-Cloud and a few miles west of Paris. The château was later bought by King Louis XIV and André Le Nôtre remodelled its gardens. Two centuries later, during the Franco-Prussian War, the Prussian army settled its headquarters in the château but a French shelling caused the burning of the building on 13 October 1870. The city of Saint-Cloud suffered serious damage during the war and, therefore, rebuilding the castle was not a priority. It was in ruins for 20 years and was ultimately demolished for safety reasons in 1892. While the park still exists, there are no visible remains of the castle; some yew trees remind visitors where it was and that its foundations are still lying underground.

In March 2006, the same year the reconstruction plans of the Tuileries Palace gained official credibility,[12] an association calling for the reconstruction of the Saint-Cloud castle (Reconstruisons Saint-Cloud) was set up. This initiative was taken by Laurent Bouvet (b. 1963), a real estate commercial agent, who trained in business law at the University Paris II-Panthéon Assas and the Institut supérieur de gestion in the 1980s. Bouvet was not a member of a historical society at that time. When he heard about the Guédelon project, he thought of doing the same in the Paris region. However, he chose not to call for the construction of a purpose-designed castle but of a castle which had long since disappeared. Bouvet found out that the Saint-Cloud castle had the most number of storeys among the numerous lost castles in the area. Adopting the funding model of Guédelon is Bouvet's true motivation; the Saint-Cloud castle is only a pretext (Bodet 36). It is no coincidence that Bouvet and Boumier chose the same funding model. Self-funding the reconstruction with the entrance fees would, in fact, be a way to prevent the waste of public money which would be the key argument to convince the French government to give its approval in principle. However, the basic problem with both projects is that they would settle on public land and on highly protected heritage sites; a public-private partnership would be necessary, yet impossible.

The reasons Laurent Bouvet now gives to justify the reconstruction are twofold and quite similar to the justification narratives of both the Versailles and Tuileries projects. The first argument is based on the assertion that the absence of the castle is confusing for visitors; even the director of the Saint-Cloud historical park, Sylvie Glaser, yet opposed to the reconstruction project, had to admit it (De la Guillonnière). The symbolic importance of the castle in French history and the topography of power until the nineteenth century is the second main argument of the supporters of the reconstruction. King Louis XIV's brother, Queen Marie-Antoinette and Napoleon were among the most famous owners and residents of the castle; the building was both the site of the coup of 18 Brumaire, in 1799, which brought General Napoleon Bonaparte to power as the first consul of France and led to the proclamation of his empire in 1802, and the place where his nephew, Louis-Napoléon Bonaparte, planned the coup d'état of 1851 which led to the reestablishment of the French empire a year later.

Yet, these arguments are now secondary compared to the economic challenge of the project. The recent statements of Laurent Bouvet focus on the fact that the reconstruction and the different activities that will be established in the castle would create hundreds of jobs and boost the local economy.[13] Local committees for tourism also support the project, betting on the fact that the castle would become one of the main attractions of the west of the Greater Paris area together with the Palace of Versailles, Frank O. Gehry's brand new Louis Vuitton Foundation and the Albert-Kahn Museum in Boulogne-Billancourt, still under renovation being supervised by architect Kengo Kuma.

The Reconstruisons Saint-Cloud association did not have a communication strategy before 2012. Laurent Bouvet has been the chairman, sponsorship seeker and communication manager of the project. He has dedicated himself full-time to lobbying politicians in order to reach the broadest consensus possible and to give evidence of public interest in the project. Parallel to this lobbying action, Bouvet has implemented an intuitive communication of influence using a wide range of media and events: the local and national press, television, website, Facebook groups and galas; 60 articles were published in six years.[14] In 2012, Bouvet decided to use the services of a communication agency to help him define a coherent communication of influence to raise awareness and indirectly affect decision makers.

The website of the association has several purposes: to account for the actions of communication of influence and lobbying, to explain the history of the castle and the content of the project, and to encourage donations. The illusion of a broad consensus across party lines on the reconstruction project was constructed first by letters of support from politicians or socialites or, in some cases, photographs showing Laurent Bouvet with influential or sometimes controversial figures such as Jean-Marie Le Pen, a citizen of the city of Saint-Cloud and also the founder and former president of the Front National, France's most powerful extreme right-wing party. Second, a press review compiled articles that evoke, to some extent, the project, including articles against the project or against 'identical reconstruction' that might seem paradoxical. The website also supplies visitors with elements for the better understanding of the history and architecture of the castle – old prints and a 3D virtual reconstruction made by architectural historian Philippe Le Pareux (b. 1974)[15] – and a heterogeneous selection of recent reconstruction cases in Europe. There is also a list of members of the honorary committee, patrons and other supporters of the project.

In the late 1990s, American historian Todd Fisher, then president of the Napoleonic Alliance and now executive director of the Napoleonic Historical Society, managed to raise significant funds to implement a reconstruction project of the Saint-Cloud castle, but failed to convince the French government. Lately, Mitchell Cantor, managing partner at the Law Offices of Mitchell Cantor in New York, got involved in Bouvet's project; he has been the chairman of the American Friends of the Chateau de Saint-Cloud since 2011[16] and has been intending to get the support of the philanthropist and Francophile community in the US, especially in New York.

The reconstruction project recently gained credibility when French architectural firm Atelier COS did a feasibility study in March 2014 (Atelier COS). In the three options that architects Fariza Mariano and Didier Beautemps suggested in that study, the common elements are a luxury hotel, shops and a public area including a gourmet restaurant, a lounge bar and various decorated rooms for special events and seminars. However, the possibility of establishing a museum only appears in two of the three options, but no details were announced as to what that museum would be about or what it would contain.

The presence of such equipment for the purposes of luxury tourism or business tourism is justified by the site's proximity to both the western districts of the French capital, richer and more attractive to the wealthy foreign tourists, and the business district of La Défense to the north. However, this is not a surprise when one knows what Atelier COS built its reputation on: the construction, at Courchevel, of one of the most exclusive resorts for winter sports in France as part of the flagship Cheval Blanc Hotels, a luxury hotel chain of the LVMH (Moët Hennessy – Louis Vuitton S.A.) group, and the rehabilitation of the prestigious Ritz Hotel and Place Vendôme in Paris, still in progress, at an estimated cost of €140 million. The construction of the first option of the architectural design for Saint-Cloud would cost about €162 million to €206 million;[17] the architects, being the main beneficiaries of Bouvet's project, would earn between €18.6 million and €23.8 million.


## CONCLUSION

These three case studies show that the phenomenon of 'identical reconstruction' of lost buildings do not meet a structured and consistent momentum in France. The paradox is that most of the recent reconstruction projects, whether or not successful, have impacted each other. For instance, the Hermione and Guédelon projects are common references for the Saint-Cloud and the Tuileries reconstruction plans. Some protagonists, such as Erik Orsenna, Michel Guyot, Jean Tulard, Laurent Bouvet, Pierre-André Lablaude and Frédéric Didier, are involved in or support at least two projects. 'Identical reconstruction' is now a commonplace phenomenon and even a growing niche market in which a few architects, most of them controversial for their uninhibited approach towards restoration and reconstruction, have tried to position themselves since a decade.

The only project that has been completed is the reconstruction of the royal gate of the Palace of Versailles that was a key element in an architectural layout thought of as a means to stress the importance and sacredness of the Sun King Louis XIV. This project arose from the direction of the castle and received official support from the National Heritage Commission and the French government.

The other two projects, which are based on the reconstruction of old royal and imperial residences burned during critical phases of the history of France, and then were demolished after years of ruins emanated from private, individual and supposedly apolitical initiatives. In the case of the Tuileries, the project is part of a revisionist trend that tries to revive the memory of Emperor Napoleon III. In the case of Saint-Cloud, the castle is only a pretext to reproduce, halfway between Paris and Versailles, the economic success of the model of the archaeological theme park of Guédelon in Burgundy. The leadership of these private projects is provided by individuals who are neither cultural heritage professionals nor historians;

they have been trying for years to raise awareness and reach the broadest consensus possible through direct lobbying and communication of influence via different popular media streams.

Among the justification narratives, some arguments are common to most of the cases: reconstruction is necessary for a better understanding of the sites, the buildings should not have been demolished or should have been rebuilt before, they were important landmarks and places of particular significance in French history, reconstruction would revive traditional building techniques and provide visitors with a true educational experience, or private and self-funding would not burden public finances.

There is no direct or explicit link with monarchist movements in their justification narratives although there are many members of princely dynasties in the committees of honour and among the supporters of the Saint-Cloud and Tuileries projects. Yet, the three projects developed when the main right party was in power in the first half of the past decade. They all tried to revive the memory of past monarchic regimes of historical figures, such as King Louis XIV in Versailles, Emperor Napoléon III in the Tuileries and Marie-Antoinette and Napoléon in Saint-Cloud.

## BIBLIOGRAPHY

Atelier COS. *La Reconstruction du Château Saint-Cloud*. 25 Mar. 2014. Web. 12 Jan. 2015. <http://www.reconstruisonssaintcloud.fr/Etude_Architecturale_Atelier_Cos.pdf>.

Bastoen, Julien. 'La Résurrection des Tuileries, ou la Tentation de l'Hyperréalité'. *Criticat* 5 (2010): 36–49. Print.

——. '*Faux et Usages du Faux. Quand le Clonage Architectural Redécouvre ses Origines: le Cas du Palais des Tuileries à Paris*'. *Reflecting on and Practicing the Spirit of Place / Penser et Pratiquer l'Esprit du Lieu*. Quebec City: Presses de l'Université Laval, 2011. 71–84. Print.

Blin, Sylvie. 'Fallait-il Reconstruire la Grille de Versailles?' *Connaissance des Arts* October (2007): 8. Print.

Bodet, Marie-Laetitia. *Le Positionnement. Etape Indispensable de la Conception d'un Projet de Communication d'Influence*. Université Est Créteil, graduate thesis, 2012.

Boumier, Alain. 'Faut-il reconstruire les Tuileries?' *Académie des Beaux-arts*. 26 Feb. 2003. Web. <http://www.academie-des-beaux-arts.fr/actualites/travaux/Boumier.pdf>.

Brown, David. 'Des Faux Authentiques. Tourisme Versus Pèlerinage'. *Terrain* 33 (1999): 41–56. Print.

Cabestan, Jean-François. 'Château de Versailles. La Reconquête du Public'. *AMC* June–July (2008): 53–6. Print.

Dawans, Stéphane and Claudine Houbart. 'Identical Reconstruction and Heritage Authenticity'. *S.A.V.E. Heritage Conference. IXth International Forum of Studies 'Le Vie dei Mercanti'*. Naples-Capri, 9–11 June 2011. Unpublished.

'Débat sur les Questions d'Achèvement, de Restitution et de Reconstruction d'Edifices'. *Monumental* 1 (2010): 100–107. Print.

De la Guillonnière, Aymard. 'Saint-Cloud veut Renaître de ses Cendres'. *LePoint.fr.* 20 Jun. 2012. Web. 29 Dec. 2014. <http://www.lepoint.fr/culture/saint-cloud-veut-renaitre-de-ses-cendres-20-06-2012-1475753_3.php>.

De Nolhac, Pierre. 'Versailles Truqué'. *Le Figaro.* 30 Oct. 1934. Print.

Didier, Frédéric. 'Le Rétablissement de la Grille Royale'. *Monumental* 2 (2005): 10–13. Print.

Duhem, Gilles. 'Château de Berlin: le Triomphe Amer de l'Architecture Rétrospective'. *Archiscopie* 26 (2002): 22–3. Print.

Ferrand, Franck. 'Les Chantiers du Château de Versailles'. *Détente et Patrimoine.* Europe 1 Radio. 10 Aug. 2008.

Gady, Alexandre. 'Faut-il Reconstruire les Tuileries? Deux Colloques pour un Etrange Débat'. *Archiscopie* 75 (2008): 22–3. Print.

Goven, François and Judith Kagan. 'La Question de la Terminologie'. *Monumental* 1 (2010): 26–8. Print.

Hasquenoph, Bernard. 'Grille en Stuc pour un Versailles en Toc'. *Louvre Pour Tous.* 26 Aug. 2008. Web. 29 Dec. 2014 <http://www.louvrepourtous.fr/Grille-en-stuc-pour-un-Versailles,077.html>.

Hernández Martínez, Ascensión. *La Clonación Arquitectónica.* Madrid: Siruela, 2007. Print.

Jokilehto, Jukka. 'Considerations on Authenticity and Integrity in World Heritage Context'. *City & Time* 2.1 (2006). Web. 29 Dec. 2014 <http://www.ceci-br.org/novo/revista/docs2006/CT-2006-44.pdf>.

Le Naire, Olivier. 'Le Château Côté Cour … '. *lexpress.fr.* 27 Sep. 2007. Web. 29 Dec. 2014 <http://www.lexpress.fr/region/le-chateau-cote-cour_474291.html>.

Levi, Daniel J. 'Does History Matter? Perceptions and Attitudes toward Fake Historic Architecture and Historic Preservation'. *Journal of Architectural and Planning Research* 22.2 (2005): 148–59. Print.

Levine, Neil. 'Building the Unbuilt: Authenticity and the Archive'. *Journal of the Society of Architectural Historians* 67.1 (2008): 14–17. Print.

Meyer, Daniel. *Les Trésors de Versailles.* Geneva: Famot, 1978. Print.

Mignot, Claude. 'Rebâtir les Tuileries? Une Lubie Sotte et Ruineuse'. *Momus* 20 (2006): 4–5. Print.

Pallot-Frossard, Isabelle. '*La Reconstruction de Monuments Disparus. Introduction*'. *Monumental* 1 (2010): 86–7. Print.

Rykner, Didier. 'Domaine de Versailles, ou Versailles-land'. *La Tribune de l'Art.* 25 Mar. 2007. Web. 29 Dec. 2014 <http://www.latribunedelart.com/domaine-de-versailles-ou-versailles-land>.

——. 'Inauguration of the "Grille Royale" at Versailles'. *The Art Tribune.* 5 Jul. 2008. Web. 29 Dec. 2014 <http://www.thearttribune.com/Inauguration-of-the-Grille-Royale.html>.

Varnedoe, Kirk. 'The Tuileries Museum and the Uses of Art History in the Early Third Republic'. *Saloni, Gallerie, Musei e Loro Influenza Sullo Sviluppo dell'arte dei Secoli XIX e XX.* Ed. Francis Haskell. Bologna: CLUEB, 1981. 63–8. Print.

Voisin, Chloé. 'Le Centre, la Mémoire, l'Identité. Des Usages de l'Histoire dans la (Re)-Construction du Nouveau Marché de Dresde'. *Espaces et Societies* 130 (2007): 87–101. Print.

Von Buttlar, Adrian. 'Berlin's Castle Versus Palace'. *Future Anterior* IV.1 (2007): 13–29. Print.

## NOTES

1   'Lo cierto es que si observamos la proliferación de clones arquitectónicos en los
    últimos años, tanto de edificios históricos como de construcciones clave en la
    arquitectura del siglo XX, deberemos llegar a la conclusión de que no es algo
    fragmentario ni episódico, sino que construye un verdadero movimiento cuyas
    motivaciones deberíamos conocer'.

2   'Lorsque l'on en vient à reconstruire à partir de rien, lorsque le temps est passé – plus
    d'un demi-siècle voire davantage – les motivations profondes et les modalités de
    mises en œuvre diffèrent largement'.

3   *Vue du Château de Versailles en 1668* by Pierre Patel, oil on canvas, 115 × 161cm, 1668,
    Palace of Versailles; *Vue du Château de Versailles Prise Depuis la Place d'Armes* by Pierre-
    Denis Martin, oil on canvas, 139 × 150cm, 1722, Palace of Versailles.

4   This replica was built between 2008 and 2011 on the basis of an 'original' idea by
    property developer Emad Khashoggi.

5   'Cette restitution s'appuie sur des bases solides et incontestables'.

6   'Nous avons passé des heures ensemble [avec les ferronniers] à mettre au point
    chaque détail, parfois d'après les archives bien entendu, mais surtout – les archives ne
    disent pas tout – par analogie à partir d'éléments authentiques conservés soit sur le
    site de Versailles soit dans d'autres grilles du XVIIe siècle'.

7   See 'Notre mission'. Web. 15 Jan. 2015. <http://www.tuileries.org/>.

8   'Un colloque pour les Tuileries', Amphithéâtre Descartes, Sorbonne, Paris,
    23 Sept. 2008.

9   Born 1947.

10  See 'Notre mission'. Web. 15 Jan. 2015. <http://www.tuileries.org/>.

11  According to the feasibility study ('prospectives'). Web. 15 Jan. 2015.
    <http://www.tuileries.org>.

12  See 'Reconstruction du Château deSaint-Cloud'. Web. 12 Jan. 2015.
    <http://www.reconstruisonssaintcloud.fr>.

13  See 'Document de synthèse'. Web. 15 Jan. 2015. <http://www.reconstruisonssaintcloud.fr>.

14  For instance, Chevalier, Justine. 'Ils rêvent toujours de rebâtir Saint-Cloud'. *Le Parisien*
    14 Aug. 2010. Web. 15 Jan. 2015. Guénot, Hervé. 'Ils veulent reconstruire Saint-Cloud'.
    *Journal du dimanche* 30 Apr. 2011. Print. Liffran, Hervé. 'L'ultime fantasme des derniers
    aristos. Reconstruire leurs châteaux!' *Le Canard Enchaîné* 8 Aug. 2012. Print.

15  Philippe Le Pareux also made a 3D virtual reconstruction of the rooms of the
    Versailles Palace.

16  Mitchell Cantor is also a member of the Board of the American Friends of the Chateau
    de Compiegne, and was a member of the American Friends of the Louvre.

17  According to a study conducted by construction consultancy Gleeds France in 2014.
    Web. 12 Jan. 2015. <http://www.reconstruisonssaintcloud.fr/Gleeds.pdf>.

# 6

# Refracted Copies of the Imperial City and the Great Audience Hall in East Asia

*Alice Y. Tseng*

Japan's adoption of Chinese imperial city planning and imperial style architecture was part and parcel of its comprehensive absorption of a well-developed foreign paradigm for its own state formation. Assimilating China's writing system and legal codes, as well as Chinese-inflected Buddhism, made possible a centralized bureaucracy and religion with a defined distribution and hierarchy of governance. As physical expression of this orderly rule, the capital correspondingly featured a perfect rectangular outer boundary with orthogonally configured city wards; an outsized top centre sector defined a privileged location for the palace city. Despite inevitable differences in scale, proportion, and topographical conditions, the early Japanese imperial capitals have been commonly described as copies of Chinese imperial capitals for their adherence to the same planning principles. The notion of copying is reinforced by the existence of Japanese multiples – Naniwa, Fujiwara, Heijō, Kuni, Nagaoka and Heian, dating from the seventh to the eighth centuries – that successively repeated the original Chinese formula for capital making.[1]

An analysis of the aforementioned Japanese capitals individually or as a collective type therefore begs consideration of acts of cross-cultural repetition, replication, and reproduction as stimuli to state building. Furthermore, acknowledging the sequential development of these Japanese capitals, each a unique interpretation of the Chinese prototype, triggers questions regarding origin and deviation in evaluating the reproduced form as historically significant. The imperial city plan that crossed culture and time assumes yet another layer of interpretative complication when these ancient sites are reanimated in the twentieth and twenty-first centuries by archaeological and architectural reconstruction in situ. In short, reproduction in its broadest possible sense defines the very origin, recurrent validation, and modern memory of the ancient Japanese imperial capital. Heijō, the capital from 710 to 784, is perhaps the richest embodiment of such a multiplicity of reproductive efforts. Not only did it represent the climax of a fertile spurt of Japanese copies of a Chinese paradigm – as it was realized at the city of Chang'an – the main palace section of Heijō has been the object of archaeological surveys and preservation efforts continuously since the mid-nineteenth century. Heijō would experience a renewal of kinship with Chang'an in the late twentieth century when both sites saw the completion of a major phase of restoration and the reconstruction of main structures, facilitating an updated understanding of their similar, yet far from identical, arrangement and appearance. To investigate Heijō, then, is to understand the elasticity and potency of reproduction, capable of generating profound cultural-historical meaning, notwithstanding certain poverty of originality.

This chapter focuses on the re-presentation of Heijō in the early twenty-first century by investigating its current manifestation within the modern city of Nara. Specifically, I focus on the recovery of the palace section that was accomplished through a sustained, nationally-directed effort in excavation and reconstruction (Figure 6.1). I make close reference to a separate but ideologically parallel project of cultural-historical recovery at Xi'an, China, developing concurrently. The year 2010 witnessed the unveiling of both projects to the public. In Xi'an, the Daming Palace Heritage Park (Daminggong Guojia Yizhi Gongyuan) opened on the excavated site of the imperial city Chang'an from the Tang period (618–907). That same year, the city of Nara celebrated its 1300th foundation anniversary on the excavated site of the Heijō Palace (Heijō Kyūseki) from the Nara period (710–94). At both sites, more than five decades of archaeological excavations were expended for the recovery of the earthen foundations of the original palace buildings. To accompany the 2010 unveilings, the Daming Palace Heritage Park and Heijō Palace Site both opened on-site museums that display unearthed relics and discovery centres that provide interactive activities to facilitate understanding of the past. The reclamation of antiquity was taken to the next step with the full-scale reconstruction of major structures. The reconstructions provide tangible physical forms and fixed images of the long-lost ancient structures, belying the host of educated conjectures developed to generate the architectural form and details.

These literally groundbreaking projects in China and Japan, distinguished by the enormity of physical scale, intellectual and physical labour, and supported by national government funding and UNESCO sanction, have received very little media attention abroad. Nor have they been examined critically as possibly problematic examples of architectural replication. The credibility of a nation's reproduction of its own historical past has not attracted attention or derision the same way that Chinese and Japanese cross-continental, cross-temporal copying of the Eiffel Tower or whole European villages has.[2] No doubt the protection of national heritage cleanly trumps the promotion of leisure industry interests as a worthy purpose of architectural reproduction. Yet reconstruction, even with the best of intentions for preserving authenticity and achieving accuracy, remains controversial as an ideal course of action for engagement with the past (Jameson 1–14). The significant historical value of the two ancient imperial capitals, Chang'an and Heijō, is not in dispute here, nor is the recognition that the cities' extant remains are of 'outstanding universal value', according to UNESCO's 1972 World Heritage Convention (2). I am instead focusing on the nascent reconstructions as the latest chapter in a longer history of copying that has shaped the Sino-Japanese imperial city in concept and form.

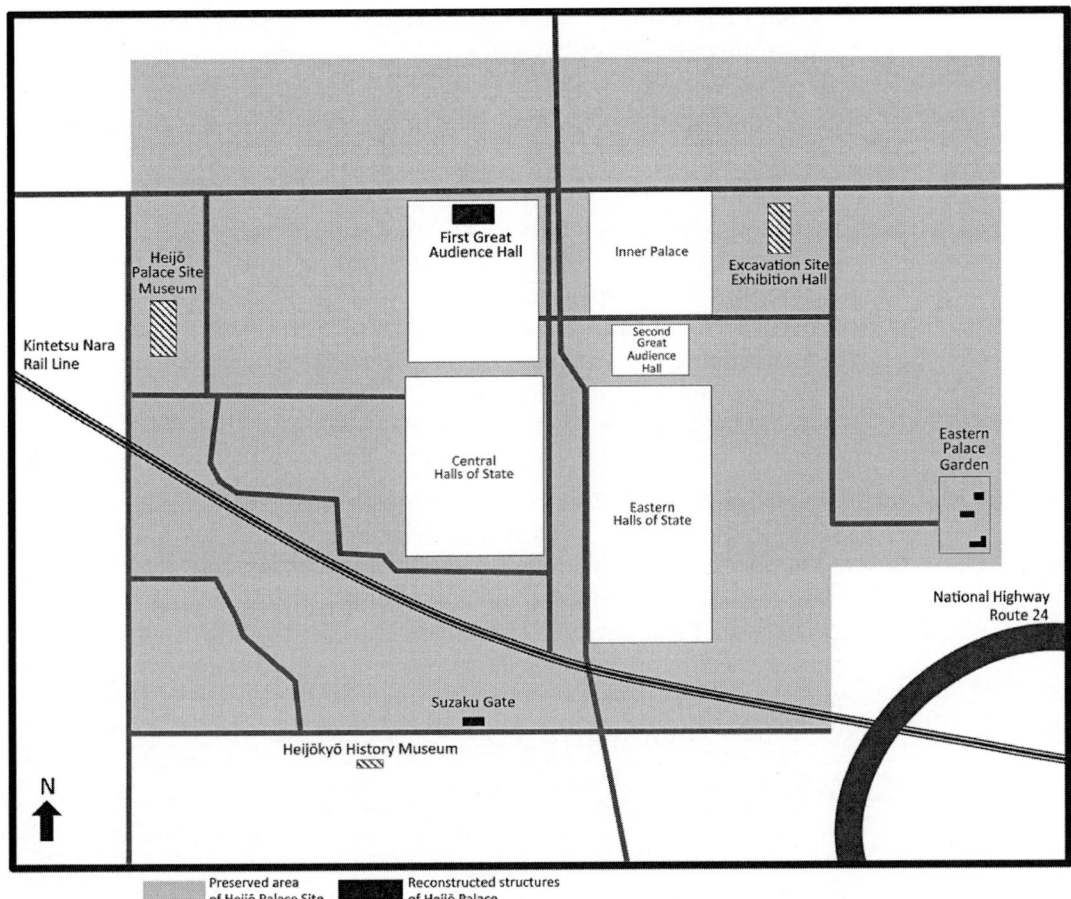

The replicas in this instance support the ideology of heritage preservation and dissemination for the recollection of a distant past palpably disconnected from the present. Meeting the challenge of vivifying a foreign past gave rise to two problematic approaches that I will examine: the erection of architectural copies of indeterminate accuracy on active archaeological sites, and the conversion of these same archaeological sites into museum-parks for the general public.

6.1   Heijō Palace Site, at the time of the 1300th foundation anniversary, Nara, Japan, 2014
*Source:* Alice Y. Tseng.

## AN INTERLINKED BEGINNING

Chang'an's Daming Palace and Heijō's Imperial Palace were roughly contemporaneous creations, being founded within 50 years of each other.[3] To begin exploring the linkage between them, however, is to confront a labyrinthine course between the presumed 'source' and 'derivative': the original turns out not precisely novel and the copy not so faithful an imitation. While Chang'an served as the original to Heijō, the city itself represented a kind of copy, a posthumous homage, to a much earlier Chang'an of the Han dynasty (206 BCE–220 CE). The founding emperor of Tang revived the name, supplanting the existing name Daxing it had assumed under the preceding Sui dynasty. Palaces and halls from the old Chang'an were transferred to the new, located a short distance of ten kilometres southeast from its namesake (Wu 35). Although roughly rectangular, Han Chang'an did not yet feature the regularized gridiron layout of streets or the north-central location of the imperial city and palace city, which were characteristics of the prototypical Chinese imperial city established at Daxing and finalized during its manifestation as Chang'an (Steinhardt 1990 94–108). Concurrently, Daxing had spawned a copy at a smaller scale, a secondary capital at Luoyang, the replication facilitated by one designer, the imperial architect Yuwen Kai (555–612), who participated in the planning of both cities (Steinhardt 1990 96). The name Luoyang, too, was revived to reference an original from the Han dynasty. While the older Luoyang approximated the geometric four-sided enclosure, the newer Luoyang, due to an unplanned western portion, actually lost the defining symmetry that had begun to take shape at the Han city six centuries earlier and was perfected at the recent Daxing. The second Luoyang seemed to reflect aspects of both the first Luoyang and Daxing without charting a clear linearity of development of one singular archetype.

The grandeur and sophistication of the Tang Chang'an, a city of 1,000,000 in population that witnessed the height of ancient China's international trade and diplomacy, exerted palpable influence over its eastern neighbours, Korea and Japan, who followed closely Chinese modes of administration, planning and construction. Just as the cases of Chang'an and Luoyang exhibited multiple manners of copying, the transference of the imperial city archetype from Chang'an to Heijō was mediated by other attempts at replication along the way. The city of Fujiwara (694–710) preceded Heijō as the first Japanese capital to assume the ideal Chinese rectangular form, symmetrical and gridded, and erect a main audience hall (daigokuden) in the palace in Chinese-style materials and form. Both Fujiwara and Heijō assumed only a fraction of the enormous size of Tang Chang'an, which measured 84 square kilometres compared to Heijō's 24 square kilometres. They also did not adopt a number of defining features such as a continuous wall around the city perimeter and high walls around individual wards. Furthermore, while Fujiwara adhered to the geometric purity of the rectangular imperial city, Heijō from the start accommodated a prominent eastern bulge known as the 'Outer Capital' (Gekyō) beyond the perfect rectangular boundary to house the prominent Buddhist temples Kōfukuji and Gangōji (Shirai 12). Although irregularity could be found at Tang Chang'an as well, the trapezoid-shaped protrusion from the northern edge of the city represented the later addition of a summer palace that eventually turned into the Daming Palace. Despite the disrupted geometry and symmetry, this imperfect city form would be what the Japanese envoys perceived as Chang'an when as many as 18 official embassies were sent from the mid-seventh to the ninth centuries.

Beyond records of Japanese embassies having taken place, scholars have yet to determine the exact method or condition of knowledge transfer of Chinese imperial planning to Japan. While the establishment of official relations between China and Japan enabled direct contact, notably visits to the great capital of

Chang'an (Wang), Japan in the pre-Tang period also relied on the Korean kingdoms for relaying of Chinese construction technology and architectural style, especially for monumental religious and state structures. Given that multiple Chinese capitals embodied the characteristics of the ideal imperial city, and that multiple Korean and Japanese cities exhibited strong Chinese planning influence, there is yet to be a definitive answer as to whether Tang Chang'an was the immediate and sole source of inspiration for Heijō (Fu 130–32).

My short and admittedly grossly simplified account of reproduction and the East Asian imperial capital should be sufficient to frustrate any insistence on the primacy of originals over copies or of precision in replication. When little appears to be known about the historical process under which the copying occurred, and the produced copies exhibited enough latitude to suggest incomplete knowledge or intentional customization, our understanding of these cities as case studies of influence and transmission is still partial at best.[4]

## TWENTIETH-CENTURY RECOVERIES

One of the biggest obstacles to contemporary comprehension of the interrelationship among the multiplicity of imperial cities, Chinese, Korean and Japanese, created between the sixth and eighth centuries has been their physical disappearance. Chang'an and Heijō suffered similar fates after they stopped functioning as imperial capitals in the tenth and ninth centuries respectively. The palace sites deteriorated into rice paddies and untended wilderness that progressively buried the sophisticated network of wide avenues and assembly of lofty, polychromatic architecture. It was not until the 1950s that official excavations began in earnest at both places, work that has been ongoing up to today.

The ruins of Heijō, located in current-day Nara city, had attracted archaeological and architectural interest since the beginning of the modern period around the 1850s. In particular, the palace city in the north centre sector, where the central administrative offices, state ceremonial halls, and imperial residential quarters assembled, consistently received the attention of surveys and preservation efforts. Regional officials and administrators were the first to inspect the ruins and generate drawings of palace plans, starting with Kitaura Sadamasa who produced the *Heijō Inner Palace Site Zoning Map* (Heijōkyū daidairi ato tsubowari no zu) in 1852, and Tanada Kajūrō who expanded upon the former in the *Heijō Inner Palace Area Map* (Heijō daidairi shikichi zu) in 1900. At this time, the architect and pioneer surveyor-cum-historian of East Asian architecture Sekino Tadashi (1868–1935) discovered the location of the Great Audience Hall, the principal ceremonial structure of the palace, while working as a technical expert for Nara Prefecture. In 1907, Sekino would publicize his findings in an academic journal published by his alma mater, Tokyo Imperial University, the institution that since its founding in 1877 functioned in essence as the intellectual arm of the nation. Thereafter, the movement to mark and preserve the Great Audience Hall site gained momentum in the 1920s, culminating in its designation as a historic site. Activities experienced a renewed surge in the 1950s with the national government's establishment of the Nara National Research Institute for Cultural Properties (Nara Kokuritsu Bunkazai Kenkyūjō), under which the Division of Heijō Palace Site Investigations was established in 1963 to oversee excavation and research of this site. During this decade the national government purchased the land that comprised the original palace complex, to stem any further loss or damage as well as prevent the incursion of a proposed national highway bypass on its eastern periphery.

In the case of Chang'an, while historiographical interest continued after the Tang period, frequently expressed through illustrated interpretations inclusive of Daming Palace, as recorded in Lu Dafang's *Plan of Tang Chang'an City* (Tang Chang'an cheng tu) of the Song period, *Illustrated History of Chang'an* (Chang'an zhi tu) of the Yuan period and *General History of Shanxi* (Shanxi tong zhi) of the Ming period (Xi'an qujiang Daming gong yi zhi qu 4), they established a lineage of pictorialization that was not supported by first-hand survey and therefore consistently inconsistent in factual details. Not until the twentieth century did the Institute of Archaeology of the Chinese Academy of Sciences (Zhongguo She Hui Ke Xue Yuan Kao Gu Yan Jiu Suo) carry out the first modern excavation from 1957 to 1960 at the site of Daming Palace, located in present-day Xi'an. Although the activities unearthed the general footprint of the full palace complex and some of its main gates and halls, funding shortage and political instability impeded continuous progress. Finally, a new phase of excavation starting in 1994 completed the survey work started earlier in the same century, and a new plan developed to direct ongoing excavation, conservation, reconstruction, and even new construction.

After more than five decades, the location of major palace gates and buildings of Daming Palace and Heijō Palace have been recovered and protected, and remaining building foundations conserved. At both sites, the focal structures are the main gate, the great audience hall and a garden. Specifically, three full-scale reconstructions were undertaken at Heijō Palace: the Suzaku Gate, which served as the main entrance to the palace city; the First Great Audience Hall, which served as the main hall where the most prominent state ceremonies were held; and the East Palace Garden, which served as the setting for the imperial family's banquets amid ponds and bridges. At Daming Palace, the full-scale reconstructions consist of the main palace gate, Danfeng Gate, and Taiye Pool, the central garden feature of the palace. The earthen foundation and elevated platform of the Great Audience Hall, Hanyuan Hall, and its flanking pavilions, have been restored, but there are no plans to reconstruct the enormous super structure. Notably, a full-scale replica of just one of the side pavilions, the 'Phoenix Tower' (Qifengge), was created for the Shanghai World Expo in 2010, the same year that Daming Palace Heritage Park opened.

The aforementioned reconstructions and restorations are accompanied by history and archaeology museums onsite that offer general information about the historical periods of Tang and Nara, respectively, and exhibitions of scale models and excavated artifacts. Newly reorganized as the Daming Palace National Heritage Park and the Heijō Palace Site, what once were active excavation sites for expert archaeological teams opened to the general public in 2010. If the timeline and approach to the recovery, conservation, restoration, and reconstruction at

Daming Palace and Heijō Palace, two distinct sites in two different countries, seem strangely uniform, it is not entirely coincidental.

From the time of Sekino Tadashi up to the end of the Second World War in 1945, surveys of ancient art, architecture and archaeological remains sponsored by the Japanese imperial government served to narrate a linked cultural past among Japan, Korea and China (Pai). Rather than limit research and excavation activities within national boundaries, leading intellectuals of the first half of the twentieth century – many of whom were on the faculty of Japan's network of Imperial Universities – actively crossed over to the continent to formulate an inclusive cultural entity known as East Asia (Tōyō). The treasures, buildings and ruins they identified would provide the material and scientific evidence of kinship that justified ongoing Japanese encroachment in the search for its continental origins. The Japanese vested interest in promoting Chinese and Korean heritage, originally cultivated and enabled by imperialist expansionism, implanted an ongoing archaeological attentiveness to an East Asian cultural nexus even after 1945.

Starting in 1994, the conservation of Hanyuan Hall became more than a Chinese national project. The UNESCO / Japanese Trust Fund for the Preservation of World Cultural Heritage enabled the Japanese government to provide subsidies and expert consultation to the restoration work here, among a handful of other prominent heritage sites.[5] A committee of Japanese and Chinese representatives in collaboration with UNESCO formulated the guidelines for the Hanyuan Hall project. In addition, throughout the 1990s and early 2000s, the respective expert teams in charge of the two imperial sites, Japan's Nara National Research Institute for Cultural Properties, and China's Institute of Archaeology's Tang Fieldwork Team, collaborated on research and excavation projects on their sites (IA CASS). In one official UNESCO document jointly authored by the Chinese and Japanese governments, the significance of Daming Palace is described as follows:

> The Daming Palace was the largest palace in Chang'an during the Tang period. When Chang'an was at its peak, the Daming Palace was the nerve center of Tang politics. Its main hall was the Hanyuan Hall where various state ceremonies were conducted. It was frequently the stage for diplomatic exchange as well. The surviving ruins of the Daming Palace thus represent an integral part in the cultural heritage of not only China, but the entire world as well. (Japan Ministry of Foreign Affairs 2, emphasis by author)

While certainly of significance to world culture, Chang'an directly affected Japan's cultural heritage by literally giving shape to the latter's ancient imperial city planning and architecture. Heijō derived its international significance in large from being a copy, or at least a copy of the concept, of Chang'an. For the two nations and two archaeological teams to collaborate, and for Japan to provide funds to preserve the physical site of Chang'an and in turn solidify this lost city's place in world history, without doubt simultaneously heightens the profile of Heijō as a collateral city. One can also argue that the current collaboration mirrors the original spirit of Tang Chang'an, generally considered the eastern terminus of the Silk Road,[6] as an epicentre of cultural and commodity exchange. The emphasis of

Sino-Japanese interchange takes physical form in the key exhibits at the Heijōkyo History Museum located at the Heijō Palace Site. A full-size replica of a ship that originally carried Japanese diplomats from Heijō to Chang'an in the eighth century allows visitors to enact the experience of crossing the sea to the continent. The same museum features two theatres with panoramic screens to reproduce and dramatize in animation and virtual reality technology the reciprocal experiences of diplomatic encounter: one of foreign envoys in Heijō and the other of Japanese envoys in China.

## RECONSTRUCTION WITHOUT AN ORIGINAL

At Heijō, as at Chang'an, only the palace section of the original city has been the focus of excavation and protection. In the final section of this chapter, I focus on the modern reconstruction of the First Great Audience Hall at the Heijō Palace Site as a prime example of architectural reconstruction as an unstable act of historical reproduction (Figure 6.2). Like the entire imperial city, this main ceremonial building is believed to have conformed closely to a Chinese architectural type as well as a specific structure. The practice of naming the main palace structure in a Chinese imperial capital the *taiji dian*, or the Hall of the Absolute Supreme, has been traced as far back as the Three Kingdoms period in the third century. Tang-period Chang'an also featured a Hall of the Absolute Supreme as the foremost structure in the north–south axis along which main buildings aligned in the Palace City, named the Palace of the Absolute Supreme (Taiji Gong). 'Daigokuden' represents the Japanese reading for the same Chinese name, with the noted exception that the first character of the three-character name had been changed inexplicably from 'tai' (greatest) to 'dai' (large). *Nihon Shoki*, the second oldest known written history of Japan, contains the first mention of a 'daigokuden' in the imperial palace at Asuka (capital city from 538–710) in the mid-seventh century. Following Chinese practice, Japanese continued to designate the main state hall the 'daigokuden' in successive capital cities Fujiwara, Heijō, and Heian (794–1868) (Zhongguo she hui ke xue yuan 428–31).

Onsite at Heijō Palace are traces of two successive Great Audience Hall structures. The earlier structure, completed in 715, had been dismantled and moved to the short-lived imperial capital Kuni (740–44) (Ooms 80).[7] When the imperial court returned to Heijō, it rebuilt a full suite of palatial architecture set to the east of the first site, albeit with a new hall realized on a smaller scale, measuring roughly 80 to 85 per cent of the first. The building chosen for reconstruction in the twenty-first century is the earlier one, and it has been labelled the First Great Audience Hall (Daiichiji Daigokuden) to differentiate from the later, Second Great Audience Hall (Dainiji Daigokuden) that dated to the second half of the eighth century.

Hanyuan Hall, which functioned as the Great Audience Hall of Daming Palace, has been proposed as the original to the Japanese First Great Audience Hall (Coaldrake 62). Even though the Hall of the Absolute Supreme of the Palace of the Absolute Supreme within Chang'an city proper was also a likely model, not enough evidence

has been unearthed at the site to provide a definitive association.[8] Instead, the archaeological teams have characterized the First Great Audience Hall as a near exact copy of the Hanyuan Hall because of the identical intercolumnial span found at the two buildings, as if they worked off of the same plan. However, the precision of the copying fails easy categorization. As the Japanese capital assumed only one-fourth the size of the Chinese original, a truncated version of the Hanyuan Hall plan was implemented at the smaller Heijō site. Besides the discrepancy in footprint area, the buildings differed considerably in overall height and formal configuration. According to the Daming excavation, Hanyuan Hall

> *was raised on a high mound, more than ten meters aboveground at its front [ … ] the eleven-by four-bay hall was [enclosed] by a one-bay-wide veranda on all sides. It projected a simple hipped roof [ … ] covered arcades stretched eleven bay-lengths across the front and back sides and four bay-lengths in depth [ … ] leading to triple-bodied pavilions [named Xianluan Tower and Qifeng Tower]. (Fu 101–2).*

The First Great Audience Hall, in comparison, was a nine-by-four-bay hall, without additional veranda. The structure was raised on a substantially lower, 3.4 metre, podium, and it stood alone, without connecting towers (Nara Bunkazai Kenkyūjo 2010b 6).[9]

6.2   The First Great Audience Hall, Heijō Palace Site, reconstruction completed in 2010 *Source:* Alice Y. Tseng.

The dramatic, three-lane stepped approach to the towering Hanyuan Hall, known as the 'dragon-tail path' (*longwei dao*), is believed to have been reproduced at Heijō rather differently, in the form of a wide, horizontal platform instead of the distinctly defined stairways.

Substantially more challenging than determining the ground-level configuration of the First Great Audience Hall was establishing the physical form of the original super structure. Even the meticulous and protracted excavation and investigation of the site by the Nara National Research Institute for Cultural Properties could not provide much information about a wooden building that had remained in place for less than 30 years and that had been completely disassembled and moved to another city. The architectural reconstruction had to be reliant upon a bricolage of indirect sources of information, models and comparisons. With no surviving buildings from the Heijō Palace nor documentation of either Great Audience Hall, textual or pictorial, extant from the period, researchers decided to consult comparable examples of monumental architecture from the seventh and eighth century for every specification of the structural composition and form of the First Great Audience Hall (Shimada 31–2; Shimizu 4–7).

Two major Buddhist structures located near the Heijō capital served as models. Having decided that the First Great Audience Hall was a multi-storey construction with a hip-and-gable roof, the research team worked to determine the type of structural system achievable in the ancient period. The immense overall scale, multiple layers of roofing, and rectangular plan of the building would have been enormously challenging to execute in timber construction,[10] and the only comparable example still extant was the renowned Golden Hall of Hōryūji temple (dated late seventh century or early eighth century).[11] Although smaller and more compact, the Golden Hall features a false upper storey on the interior to achieve the two-storey elevation visible on the outside. The same structural system was adopted for the First Great Audience Hall.[12] To realize the building's elevations and sections, the East Pagoda of the Yakushiji (dated 730) supplied the standard for the system of proportions and myriad components that form the entablature, columns and supporting bracket sets.

Given that both the Hōryūji Golden Hall and Yakushiji East Pagoda are examples of religious, not palatial, architecture from the early eighth century, the precise applicability of their form and structure for monumental palace building cannot be substantiated. Shimizu Shigeatsu, a member of the Nara National Research Institute expert team, explicates the selection of reference buildings as a matter of consulting ancient architecture (*kodai kenchiku*), apparently without regard for the incongruity of building type (6). To be fair, in view of the dearth of extant wooden buildings in China, Korea and Japan, a more comparable source does not exist. The Golden Hall and East Pagoda exemplify the world's earliest wooden architecture extant, predating even the oldest standing wooden buildings in China or Korea today.[13] The lack of a wider range of typological examples available has obliged scholars to assume Buddhist style as the blanket style for monumental public architecture of the same period.[14]

Illustrated works also played a formative role in conjuring the details of the building. A major source of information was the *Nenjū gyōji emaki* (Picture Scroll of Annual Events), a large set of paintings created at the end of the twelfth century that documented the rites and ceremonies of the imperial court. In the modern period, only the seventeenth-century reproductions of a small portion of the original scrolls survive.[15] Another twelfth-century set of paintings consulted was the *Ban Dainagon ekotoba* (Scroll of the Story of the Courtier Ban Dainagon), which illustrates a tale of intrigue at the imperial palace. Because neither artwork was intended as architectural documentation, the illustrations emphasize the human actors and their activities and offer greatly abbreviated views of the architecture. The horizontal format of Japanese narrative scrolls by convention allowed the depiction of little to nothing above the building entablature, and painters regularly applied generously-sized bands of floating clouds to the top register, further obscuring the buildings being portrayed. Although both sets of painted scrolls do offer rich visual evidence of court activities, rituals, and costumes, the city and palace illustrated in them are not Heijō, but Heian, a subsequent imperial capital that was similarly planned according to the Chinese ideal epitomized by Chang'an. As documentary evidence for the early eighth-century Heijō Palace's First Great Audience Hall, the twelfth-century scrolls depicting the Heian Palace's Great Audience Hall are untenable in almost too many ways to count.[16] Similar to the Heijō Great Audience Hall, no strong visual or textual documentation of the Heian Great Audience Hall is extant, and the assumption that the later building is a copy of the earlier has not helped to yield more reliable evidence.

In short, the reconstructed First Great Audience Hall currently standing over the excavated foundation represents a very loose interpretation, an inconclusive one, of its original form. In their publications, including those intended for the general public, the Nara National Research Institute for Cultural Properties discloses a degree of interpretative uncertainty in carrying out the reconstruction. In one brochure, in the description of the structure's roof shape, they remark:

> [The shape of the roof] was approximated based on the roof shape of the Heian Palace Daigokuden depicted in the Nenjū gyōji emaki [Picture Scroll of Annual Events] and examples of multiple-story Buddhist main halls erected before the Nara period. In the process of research, a hipped roof [yosemune tsukuri] and another type of hip-and-gable roof with defined ridges [shikorobuki] were also considered. (Nara Bunkazai Kenkyūjo 2010b n. p.)

No further explanation is supplied for favouring the specific type of hip-and-gable that they implemented. Throughout the brochure, the piecemeal approach to reconstruction is disclosed: the balustrade is based on the Hōryūji Golden Hall and the Yakushiji East Pagoda; the jewel on the central finial based on the Hōryūji Yumedono; the tiled roof ridge based on the Hōryūji Tamamushi Shrine; the pattern of the gold metal fittings based on the excavated samples from Daikandaiji; the calligraphy on the name plaque taken from characters written in the *Nagayaōgankyō Sutra* (dated 712). It should not come as a surprise that the buildings that served as models for the First Daigokuden are all registered as National Treasures of Japan, and that they represent the most methodically surveyed and researched historical architecture of not only Japan but of East Asia.

In 1998, UNESCO listed the Historic Monuments of Ancient Nara as a World Heritage Site. The list of properties included the Heijō Palace Site along with Yakushiji and four other major temples and one shrine. Hōryūji and its area temples had been inducted earlier in 1993. The work to reconstruct the First Great Audience Hall at the Heijō Palace Site began in 2001, in time for the celebration of the 1300th foundation anniversary of the Heijō capital in 2010. Despite the aforementioned assortment of conjectures and approximations that realized it, the First Great Audience Hall, raised as the central feature of the excavated palace site, does not appear tentative or improbable.

Standing like a mirage as the lone complete structure in an open archaeological field of residual foundation stones and pillar holes, it visualizes antiquity for the visitor in absolute terms. The construction team not only created a coherent exterior form but also realized it through the use of tools and carpentry methods they believed to be faithful to the eighth century. The interior walls and ceiling are composed of colours, materials, methods, and symbols meaningful to the period, as explanatory placards and brochures attentively explain. Consisting of a single open room, the building contains but one piece of furniture: the imperial throne known as the *takamikura*, topped with golden phoenixes. Wall and ceiling panels were painted by hand by the traditional-style painter and esteemed Japanese Art Academy member Uemura Atsushi (1933–) to depict Chinese and Buddhist motifs: the 12 animals of the zodiac, the four directional gods, and lotus flowers. The vermilion-painted timber members and white-washed wall surfaces on the exterior continues on the interior. From column paint colour to bracket size, gold metal patterning, to tile configuration, information about each detail is printed on information placards for the visitor to paint a comprehensive vision of a time so distant from today. The only twenty-first-century interventions are the glass panels sealing what otherwise should be an unenclosed front façade, and a wheelchair accessible ramp at the back of the building. Additionally, indiscernible to the visitor's eye would be the 'mortar-less' tiling method on the roof innovated to lighten the top load and the seismic proofing system inserted into the foundation.

The year-long celebration of the 1300th anniversary enlivened the site with activities such as ancient sports and musical concerts. Nara-period sumo wrestlers, archers, aristocrats, musicians and acrobatic entertainers populated the various steps, gardens and open areas of the palace site. Costume rentals availed to visitors enjoined them to participate in the eighth-century festival in proper period attire. School-age children could learn more about ancient history, culture, and the archaeological processes for recovering them at multiple onsite institutions: the Heijō Palace Site Museum, the Excavation Site Exhibition Hall, and the Heijōkyō History Museum. To amuse visitors of all ages, even the youngest family members, a mascot invented by the Nara City Office known by the endearing name of Sento-kun, a Buddhist-inspired cherubic boy with deer antlers, made regular appearances throughout the sprawling palace ruins.[17] Much like Mickey Mouse at Disneyland, he posed cheerfully for souvenir photos, and his image adorned an array of commercial goods, including edibles and tchotchkes that lined the gift shops and concession stands stretching from the nearest train station to the palace site. Without question, this ancient archaeological site touted its purpose as preservation, education and entertainment all rolled into one.

A similar merger of the same three pursuits has been shaping at the Daming Palace Heritage Park. Although it is beyond the scope of this chapter to pursue a parallel analysis, what deserves special mention is the additional facet of high-end commercial and residential development alongside this park. The Daming Palace Retail Cultural Complex has been placed on the southeast corner of the Daming Palace site, literally wedged between the Danfeng Gate and Hanyuan Hall. Designed to provide the first impression of the historic site for visitors after they disembark from the train station, the complex offers concentrated shopping and entertainment opportunities even before they reached the cultural–historical destination (Altoon 152–5). Nearby is a 'Garden City' development comprising five city blocks of sleek office and residential towers, luxury hotels, shopping concourses, and garden spaces. Because of the park's location within urban Xi'an, the fusion of economic and cultural interests and collaboration of government and private developer have been much more aggressive than in Nara. As part of the overall renewal of the city of Xi'an, the opening of the park in 2010 has already generated impressively high numbers of visitors, comparable to those to the internationally renowned Mausoleum of the First Qin Emperor, located approximately 30 kilometres to the east (Forte 503). At the time of writing, the Chinese government is preparing a nomination of properties along the ancient Silk Road, inclusive of the Daming Palace site, to the UNESCO World Heritage List; if successful, the influx of international commercial sponsorship and tourist traffic will surely surge, further diluting the mission of historic preservation.

## CONCLUSION

When Heijō was created in 710, it exemplified an overseas copy of Chinese Chang'an, clearly not absolutely identical, but with enough physical and conceptual correspondence for the derivation to be unmistakable. Instead of accepting the notion of one as a copy of the other as a factual endpoint, historians should consider it as the beginning point of an inquiry into the hermeneutics of copying and reproduction: to what end and through what process? The same two questions should apply to reconstruction of a lost building, so that the act of deliberate replication is investigated for its purpose and tactics of repossessing the past. The First Great Audience Hall of Heijō Palace provides much rich terrain for investigating architectural reproduction because of its historical place in the cross-continental, cross-temporal lineage of imperial Great Audience Halls of the sixth to eighth centuries and its current standing in the nexus of reconstruction projects targeted at ancient imperial sites that enmesh often contradictory priorities of heritage management and commercially-driven educational entertainment.

In attempting to achieve a modern renewal of a shared ancient culture, the roughly contemporaneous reconstruction projects at the Daming Palace and the Heijō Palace have inadvertently brought out discrepancies regarding their presumed formal likeness. Despite arguably analogous reconstructed floor plans, a comparison of the reconstruction drawings of the superstructures indicates major variation, at least to the eyes of an architectural historian. The Chinese and Japanese archaeological teams have re-created noticeably distinct characteristics, from the shape of the roof, to the width to height ratio of the main hall, to the configuration of buildings in the larger palace complex. Why then have these differences not been discussed as much as their similarities? Is it because archaeologists, who are more concerned with ground-level rather than ground-up properties of the buildings, have been the experts in charge of the research on Chang'an and Heijō? Or are factors and priorities outside the disciplinary predilections and ambitions of archaeology and architecture history steering the reconstruction efforts?

Onsite information placards written in four languages (Japanese, English, Chinese and Korean) at the Heijō Palace Site at first glance seem forthcoming with the mission and methods of reconstruction, although if the visitor reads them with some care, they also divulge a paradoxical stance on authenticity and accuracy. An initial printed greeting sets up the project as a way to interpret the reality of a lost past: 'the aim of [the reconstruction of the Former Imperial Audience Hall] was to maintain the legacy of traditional building techniques, while providing visitors a place to experience the Nara Palace as it stood in the Nara period'. However, the following placard quickly deflates the premise laid by the first by disclosing: 'No data directly indicating the Former Imperial Audience Hall's appearance have survived', and that 'research was conducted on [ … ] buildings surviving from the ancient period, along with the Heian Palace Imperial Audience Hall depicted in the *Nenchū gyōji emaki* (Illustrated Scroll of Annual Events and Ceremonies) of the Heian period, to reconstruct the original appearance'. The average visitor is being invited to 'experience Nara Palace as it stood in the Nara Period' through an architectural reconstruction that admittedly relied on no shred of direct evidence.[18]

However suspect the methodology, the reconstruction at both palace sites has significantly boosted the construction of national heritage for China and Japan, respectively and collaboratively. To emphasize their similarity – Heijō as a reflection of Chang'an, the Nara period as a reflection of the Tang period – serves to mutually enforce their shared ancient culture of emperor-centric monumentality, elevating these structures as the East Asian equivalent to European classical antiquity's Seven Wonders of the Ancient World. As marvels of construction and artistry, the Sino-Japanese Great Audience Halls were not only man-made structures of spectacular size and pageantry, they each served as the pinnacle point of an ideally configured capital city founded as a fully formed expression of heavenly-mandated imperium. Yet given the political proclivity of the post-Second World War governments of China and Japan that have repudiated absolute monarchy, the restoration projects have been propelled instead by the momentum of cultural preservation, and as multi-national efforts. The collaboration of Chinese and Japanese specialists, the observation of UNESCO guidelines, and the response to public education and entertainment, have all worked to de-emphasize the historical complexity of transferring the concept of Chang'an to Heijō, and the modern complexity of restoring three-dimensional structures based on ground-level and below-ground archaeological evidence, supplemented by oblique textual and pictorial documentation.

The expediency of presenting history and heritage through architectural reconstruction cannot be disputed; the visual clarity that it affords is both its best and worst attribute. Agreeing with Groom that a copy is theoretically always 'a degradation of the original', and with Jameson that 'a true replication of the past can never be achieved', it is the assumption of the transparent copy that needs to be handled with care (Groom 9; Jameson 3). The relationship between architectural original and copy, as the Daming Palace and Heijō Palace case studies have shown, can be far from reflexive. By calling attention to the multiple trajectories of transhistorical and transcultural copying, attended by the physical imprecision of replication during each transmission that had to occur in the realization of the First Great Audience Hall at the Heijō Palace Site, I hope to have provoked a rethinking of the architectural copy as an inherently interpretive approach to a larger cultural–historical production rather than the sum of empirical operations to beget a one-to-one replica. Purists and specialists may rightfully cringe at the reconstructions at Heijō Palace as a flimsy, speculative exercise and even conclude that the reimagined First Great Audience Hall serves no useful historical purpose. Ultimately, it is the entire site's reactivation of interest in the culture and technology of an interlinked ancient world that will propel further investigation, and hopefully elucidate the many unknowns that plague early twenty-first century reconstructions.

## BIBLIOGRAPHY

Agnew, Neville. *Conservation of Ancient Sites on the Silk Road: Proceedings of the Second International Conference on the Conservation of Grotto Sites, Mogao Grottoes, Dunhuang, People's Republic of China, June 28–July 3, 2004*. Los Angeles: Getty Conservation Institute, 2010. Print.

Altoon, Ronald. *Urban Transformation: Transit Oriented Development and the Sustainable City*. Mulgrave, Vic. (Australia): Images Publishing Group, 2011. Print.

Bosker, Bianca. *Original Copies: Architectural Mimicry in Contemporary China*. Honolulu: University of Hawaii Press, 2013. Print.

Coaldrake, William H. *Architecture and Authority in Japan*. London: Routledge, 2002. Print.

Cox, Rupert, ed. *The Culture of Copying in Japan: Critical and Historical Perspectives*. London: Routledge, 2007. Print.

Forte, Maurizio. 'Virtual Worlds, Virtual Heritage and Immersive Reality: The Case of the Daming Palace at Xi'an, China'. *Handbook on the Economics of Cultural Heritage*. Eds. Ilde Rizzo and Anna Mignosa. Cheltenham: Edward Elgar, 2013. 499–507. Print.

Fu, Xinian. 'The Sui, Tang, and Five Dynasties'. *Chinese Architecture*. Ed. Nancy S. Steinhardt. New Haven: Yale University Press, 2002. 90–133. Print.

Groom, Nick. 'Original Copies; Counterfeit Forgeries'. *Critical Quarterly* 43.2 (2001): 6–18. Print.

Institute of Archaeology, Chinese Academy of Social Sciences (IA CASS). 'Japanese Scholars from Nara Visit Tang Era Chang'an Archaeological Sites'. *Institute of Archaeology, Chinese Academy of Social Sciences.* 30 Mar. 2010. Web. 14 Jan. 2014 <https://www.kaogu.net.cn/plus/search_en.php?typeid=3&q=japanese+scholars+from+nara>.

Jameson, John H., Jr., ed. *The Reconstructed Past: Reconstructions in the Public Interpretation of Archaeology and History.* Walnut Creek, CA: AltaMira Press, 2004. Print.

Japan Ministry of Foreign Affairs and China National Administration for Cultural Heritage. *Hanyuan Hall of Daming Palace.* Beijing: UNESCO Beijing Office, 1998. Print.

Johnston, Eric. 'Nara Fears 1,300th Anniversary Flop'. *The Japan Times Online.* 6 Feb. 2010. Web. 7 Sep. 2010. <http://www.japantimes.co.jp/news/2010/02/06/national/nara-fears-1300th-anniversary-flop/#.VJQsLUBCms>.

Nara Bunkazai Kenkyūjo. *Heijōkyō: Nara no miyako no matsurigoto to kurashi* [Heijō City: State Affairs and Everyday Life at the Nara Capital]. Nara: Nara Bunkazai Kenkyūjo, 2010a. Print.

——. *Heijōkyū Daiichiji Daigokuden* [The First Daigokuden of Heijō Palace]. Brochure. Nara: Nara Bunkazai Kenkyūjo, 2010b. N. p. Print.

Ooms, Herman. *Imperial Politics and Symbolics in Ancient Japan: The Tenmu Dynasty, 650–800.* Honolulu: University of Hawaii Press, 2009. Print.

Pai, Hyung Il. *Heritage Management in Korea and Japan: The Politics of Antiquity and Identity.* Seattle: University of Washington Press, 2013. Print.

Shimada Toshio. 'Daigokuden no saigen to Nihon no kodai kenchiku' [The Reconstruction of the Daigokuden and Japan's Ancient Architecture]. *Bessatsu Taiyō* 165 (2010): 28–35. Print.

Shimizu Shigeatsu. 'Shokubai to shite no fukugen kenchiku, Heijōkyū Daiichiji Daigokden no fukugen to rekishiteki kenzōbutsu' [Architectural Reconstruction as Catalyst: The Reconstruction of the First Daigokuden of Heijō Palace and Historical Building]. *NPO ki no kenchiku* 27 (2010): 4–7. Print.

Shirai, Yoko, translated and adapted. *Envisioning Heijokyo: 100 Questions and Answers about the Ancient Capital in Nara, Japan.* North Charleston, SC: CreateSpace, 2011. Print.

Steinhardt, Nancy. *Chinese Imperial City Planning.* Honolulu: University of Hawaii Press, 1990. Print.

——. 'The East Asian Architectural Canon in the Twenty-First Century'. *Asian Art History in the Twenty-First Century.* Ed. Vishaka Desai. Williamstown, MA: Sterling and Francine Art Institute, 2007. 15–39. Print.

——. 'Seeing Hōryūji Through China'. *Hōryūji Reconsidered.* Ed. Dorothy Wong. Newcastle: Cambridge Scholars, 2008. 49–97. Print.

Tanabe, Ikuo. *Heijōkyō o horu* [Excavating Heijōkyō]. Tokyo: Yoshikawa Kōbunkan, 1992. Print.

United Nations Educational, Scientific and Cultural Organization (UNESCO). *Convention Concerning the Protection of the World Cultural and Natural Heritage.* Paris: UNESCO, 1972. Web. 18 Dec. 2014 <http://whc.unesco.org/en/conventiontext/>.

Wang, Zhenping. *Ambassadors from the Islands of Immortals: China-Japan Relations in the Han-Tang Period.* Honolulu: University of Hawaii Press, 2005. Print.

Wong, Dorothy, ed. *Hōryūji Reconsidered*. Newcastle: Cambridge Scholars, 2008. Print.

Wu, Liangyong. *A Brief History of Ancient Chinese City Planning*. Kassel: Gesamthochschulbibliothek, 1986. Print.

Xi'an qujiang Daming gong yi zhi qu bao hu gai zao ban gong shi. *Daming gong guo jia yi zhi gong yuan guihua pian* [Daming Palace National Heritage Park: Planning]. Beijing: Ren min chu ban she, 2009. Print.

Zhongguo she hui ke xue yuan kao gu yan jiu suo. *Tang Daming gong yi zhi kao gu fa xian yu yan jiu* [Archaeological Excavations and Researches on the Site of the Tang Daming Palace]. Beijing: Wen wu chu ban she, 2007. Print.

## NOTES

1    For more on Chinese imperial capitals from the Sui and Tang periods and the cities in Japan and Korea influenced by them, see Fu (92–106) and Steinhardt (1990, 108–21).

2    Japan led the way in large-scale replication of European architecture after its economic recovery from the Second World War. The Tokyo Tower, modelled after the iconic Eiffel Tower, functioned as a communications antennae and was completed 1958. Huis ten Bosch, a resort theme park in the form of a Dutch town opened 1992 in Nagasaki. More recently, in the twenty-first century, multiple Eiffel Towers and a number of European villages have sprung up in China, most notably a full-scale replica of Hallstatt, Austria, in southern China. New scholarship has attempted to divert the prevailing, mostly deprecatory, characterization of contemporary East Asian architectural copying by providing historical examples contextualized within existing social and cultural practices. See Bosker (chapters 1 and 2) and Cox (introduction).

3    Depending on how one defines the beginning of this palace, Daming was first constructed as a summer palace in 634 and did not serve as the active palace city for the reigning court and emperor until 662. Heijō Palace was constructed approximately between 708 and 715.

4    The lack of clarity regarding the transmission of planning principles and architectural style and construction has not stopped archaeologists and historians from applying a circular methodology to expand current understanding of these cities and their buildings. As Fu explains: 'Because the survival rate has sometimes been better to China's northeast than in China itself, architectural historians have looked to Korea and Japan to understand aspects of Chinese construction' (132). However, this method assumes Japanese copies were to a large extent faithful to the original – an assumption that cannot be supported due to lack of sufficient evidence.

5    Two others include the Longmen Grottoes and Kumtura Caves of the Thousand Buddhas. See Agnew (37).

6    Although Japan has helped to uphold the significance of Chang'an as a centre of international exchange, it has also vocally contested the notion of Chang'an as the terminus of the Silk Road. Instead, Japan has offered Heijō as the eastern endpoint, for a large number of the objects that circulated through the trading route are still preserved in the Shōsōin, the imperial treasure house of the eighth century.

7    After the court left Kuni to return to Heijō, the Kuni palace site was reused for the Yamashirokukubun temple complex. The First Great Audience Hall was repurposed as its golden hall. Its foundation stones remain, and the site is now preserved.

8    One scholar, Wang Zong Shu, has argued that the Japanese delegations to Tang China would have had little exposure to the Daming Palace during the years immediately before and during the construction of Heijō's First Great Audience Hall due to strained political relations: a war at sea between China and Japan in 663 and then 30 years of severed diplomatic relations. According to Wang, Japanese understanding of Chang'an imperial palatial architecture would have been mostly based on the structures within the Palace of the Absolute Supreme instead of the Daming Palace (Zhongguo she hui ke xue yuan 429).

9    A four-volume report of the reconstruction of the First Great Audience was published in 2009–10 by the Nara National Research Institute for Cultural Properties under the title *Heijōkyū Daiichiji Daigokuden no fukugen ni kansuru kenkyū* (Research on the Reconstruction of the First Great Audience Hall of Heijō Palace).

10   The main challenges would have been supporting and balancing the weight of the double height and roof trusses. As a rectangular building, it also transferred load differently from a square building (most commonly assumed by Buddhist pagodas) that had four equal-length sides.

11   The generally accepted dating for the construction of the Golden Hall at Hōryūji is around 700 to 710, which would precede the Heijō First Great Audience Hall by roughly five to ten years. For the latest published scholarship in English on the art and architecture of the Hōryūji, see Wong.

12   On the exterior, a balustrade marks the division between an upper and lower storey. From the inside, there is only one functional storey. This is due to the presence of a *mokoshi*, a pent roof that encloses the building core on all four sides. The use of *mokoshi* was typical for Buddhist main halls and pagodas.

13   Steinhardt discusses the problem of relying on the extant eighth-century Japanese wooden buildings to speculate on contemporaneous Chinese and Korean architecture that are now lost (2008). She argues that the Hōryūji Golden Hall actually references, not late seventh century, but much earlier fifth- or sixth-century continental Buddhist architecture (81).

14   While palatial architecture is assumed to have inspired religious architecture in ancient China, and Japan adopted this associated expression from the continent for its official halls and Buddhist monasteries, an indigenous style and form of construction for the imperial residence (*dairi*), the emperor's private living quarters, also endured. Such was the case at Heijō as well as Heian.

15   When Emperor Go-Shirakawa commissioned the set at the end of the twelfth century, the original scrolls numbered about 60. They did not survive intact in the following centuries. The number dwindled to merely 15 by the mid-seventeenth century, when Emperor Go-Mizunoo ordered copies of the remaining scrolls.

16   A reproduction of a portion of the Heian Great Audience Hall was created in 1895 to serve as the outer sanctuary of the Heian Shrine. Its architect Itō Chūta later publicly lamented the many interpretative mistakes he made in the design due to lack of reliable information about the historical structure. For this reconstruction, Itō's team also consulted the same painted scrolls, the *Nenjū gyōji emaki* and *Ban Dainagon ekotoba*.

17   The mascot Sento-kun designed by Yabuuchi Satoshi, a professor at the Tokyo University of the Arts, has attracted a fair amount of controversy, mostly for religious and aesthetic reasons. His physical resemblance of the Buddha in child form (suggested by the urna on his forehead), albeit with deer antlers fused on his shaven head (referencing the sacred deer who roam in Nara Park), provoked

criticism of sacrilege. The other typical complaint targeted Sento-kun's appearance as creepy rather than lovable, defying conventional expectation of cute, cartoon-like characters such as Hello Kitty of Sanrio or Domo-kun of NHK that represent Japanese corporations, institutions, and many local governments today (Johnston n. p.).

18   Shimizu frankly acknowledges that rebuilding on top of the archaeological site is a thorny endeavour, although he also argues that only by actual physical reconstruction at full scale (as opposed to creating reduced-scale models or CG renderings) could experts today comprehend the mechanics of constructing in the manner of the eighth century (4–5).

# 7

# From Historical Monument to New 'Urban Spectacle': Case Study on the Great BaoEn Pagoda Reconstruction Project in Nanjing, China

*Jing Zhuge*

## INTRODUCTION

Initiated by the Nanjing Municipal Government (NMG), reconstruction of the Great BaoEn Pagoda (GBEP) started in 2001. According to official records, the Great BaoEn Temple was built during 1413–28 under the command of Emperor Yongle of the Ming Dynasty (1368–1544) to commemorate his parents. The temple had an 80-metre-high pagoda covered with coloured glaze (Figure 7.1). The pagoda was regarded as a symbol of the Nanjing city from the fifteenth to nineteenth centuries, and greatly impressed missionaries, diplomats and soldiers who came to Nanjing from Europe after the sixteenth century. In 1854, the pagoda and the temple were destroyed during the Taiping Rebellion. Over time, the site was occupied by housing, leaving only some relic sites and place names that could remind people of the magnificent royal architecture. The GBEP reconstruction project has been ongoing for more than 10 years, and support from the government and investments have kept increasing while discourses about the pagoda were constructed which, in return, affected the project.

The project is not the only case related to the representation of a historical monument. In the past 10 years, nearly 20 similar projects in Nanjing were designated as significant by the government. In almost every Chinese city, a number of historical monuments – historical architecture or sites – were or are being rebuilt, repaired or constructed as archaeological site parks. Meanwhile, Chinese cities are changing rapidly; skyscrapers are seen as representatives of state-of-the-art technology and have become the landmark of Nanjing nowadays. Our question, therefore, is this: Why are the reconstruction projects of historical monuments like the GBEP so prevalent in China today?

It is generally believed that a historical monument bears the genuine history of its site and that, by being represented, a monument will tell a true story to today's people. Furthermore, the importance attached to historical monuments indicates a return of confidence in and respect for the traditional Chinese culture.

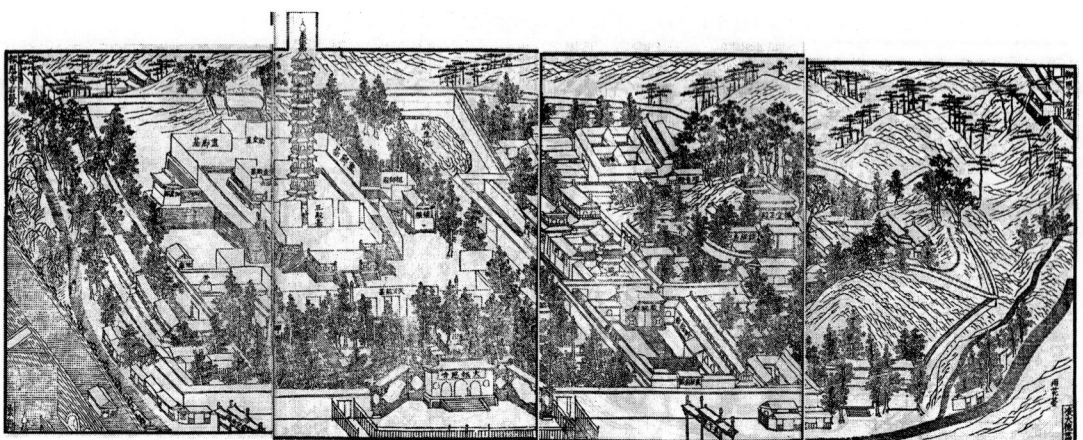

7.1   Ancient drawings of the Great BaoEn Temple from the Ming Dynasty
*Source:* Public domain. *Records of Buddhist Temples in Jinglin,* 1607.

However, there is an imagination of 'true' history implicated by these opinions. As we all know, the *truth* in history is always coterminous with the *intention* in the writing of history. These opinions conceal the divergence between different forces and discourses involved in the reconstruction programme. They also neglect the fact that destruction happens simultaneously with representation. In other words, the selection of monuments to be valued and represented is based on certain purposes. Why are only some historical monuments picked for representation? How do different forces manipulate the construction of history as well as the collective memory related to monuments? What role could the construction of a historical discourse and the representation of historical monuments play in contemporary Chinese cities? These questions are waiting to be explored. The GBEP reconstruction project is, therefore, an insightful case for discussing such questions because of its historical importance and the complexity of the reconstruction programme.

This chapter will attempt to reveal the construction process of collective memory of the history of a city by exploring the roles of the different forces in writing, advertising and designing in the case of the GBEP project. First of all, this chapter attempts to unravel how the image of the pagoda was shaped by scholars, the mass media and the public, and constructed the relationship between material and discourse, past and present, and architectures and cities. This not only justified the reconstruction and endowed the monument with immediate significance in the current context, but also pointed out that the choice of monument was closely related with the construction of the history of a city. Second, the chapter will explore the changing plans when investment and support from the government were increased and the pagoda's cultural significance was magnified in order to reveal the competition and the compromise between the various concerned parties. Finally, it will point out that the representation of historical monuments plays a significant part in the business strategy of a city in the very competitive context of today's Chinese cities. It has become a general development mode for cities in China today – driven by capital, and served by writing, advertising and designing.

## PRODUCTION OF MEANINGS

Articles on the GBEP reconstruction project have been published in newspapers and popular magazines as well as on websites from time to time since 2001. These reports were usually concerned with three topics: following up on the project progress, describing the value and significance of the pagoda over and over again and defining the project. A group of complex and comprehensive meanings of the reconstruction programme gradually emerged that were constructed and accepted by historians, architects and politicians through the mass media. An image of the GBEP was shaped by this group in the minds of the people, endowing Nanjing with a unique value. Intertwined with the history of the city, the history of the pagoda was selected, reorganized and rebuilt according to the demands of the theory constructed.

It was architectural historians who realized the pagoda's importance in the first place. They tried to recover the image of the pagoda through historical documents. Their work was encouraged by the government and improved by new archaeological discoveries. The focus of their study transformed from architectural history to Buddhist history in order to respond to the requirements of the identification of the pagoda.[1]

The first thing one would notice in the follow-up depiction is repeated citations from 'foreigners' or 'Westerners' in order to prove the pagoda's achievement in architectural history (Chen, Chen and Lu; Cai); among them, the most favourable slogan – 'one of the seven miracles in the medieval world' (Zhang Hui-Yi 5) – was due to the international perspective it lent to the pagoda. However, the citation did not tell the readers whose view it was, and a general and anonymous 'foreigner' or 'Westerner' covered the difference between the various European texts. Thus, this slogan became a kind of collective recognition, which greatly strengthened its credibility and persuasiveness because it implied that the comment was based on some universal evaluation criteria instead of a limited, local outlook. As an outsider's perspective, this comment appeared objective and neutral.

However, there was no evidence proving that most European writers believed the Great BaoEn Pagoda was one of the seven miracles in the medieval world[2] although they were truly impressed by the coloured glaze and the magnificent sight from the top of the pagoda.[3] As a matter of fact, the repeatedly cited slogan was originally from Johan Nieuhof's book of travels edited by his brother, Henry Nieuhof.[4] Chinese scholar Zhang Hui-Yi came across the book in the early twentieth century and quoted it in his *Record of the Great Bao-En Temple in Jing Lin* (first published in 1937). He added, '(the pagoda) was one of the seven miracles in [the] medieval world, famous in the world together with the Colosseum in Rome, the Tomb of Alexander and the Leaning Tower of Pisa' (Zhang 5). This statement was the genesis of the 'one of the seven miracles in the medieval world' comment that Chinese scholars and the media liberally used later.

Despite the widespread readership of Nieuhof's book in Europe, scholars noticed its unusual exaggerated style, and suspected that it was not faithful to the original manuscript (Blussé and Zhuang 14–16). In 1984, Leonard Blussé found the original manuscript of Nieuhof, which showed that the text and illustrations in the final, published work had been modified and overstated. Blussé's work, translated into Chinese and published in 1989, has been quoted by Chinese researchers. However, this did not change the mass media reports. 'One of the seven miracles in the medieval world' has almost become one of the permanent taglines for the pagoda. It seems that no one is concerned about the authenticity of the slogan.

Except the 'one of the seven miracles' from the 'Westerners', another favourable quotation about the pagoda is from *Memory of Tao An* written by Zhang Dai in the seventeenth century:[5]

> It was a great antique in China, a great porcelain of Emperor Yong-Le [ … ].
> It wouldn't be a success without the spirit, the resources, the courage and the
> intelligence of a founder of an empire. There were billions of golden Buddhist
> statues on and inside the pagoda … only God could make it. It was said that
> there were three sets of coloured glaze components when the project finished,
> one of them was used to construct the pagoda, and the other two were numbered
> and preserved … During the Yong-le time, barbarians from hundreds of other
> countries fell in full admiration and respect in front of the pagoda. They said
> nothing could be compared with the Great Bao-En Pagoda in the world.
> (Zhang Dai 18)

We can identify in this paragraph all the core factors of the discourse about the pagoda today: a symbol of an age, sophisticated technology, an architectural miracle and respect from foreigners. However, the significance of *Memory of Tao An* for Zhang Dai was the same as À la Recherche du Temps Perdu was for Marcel Proust. Both writers constructed a lost world in memory with words (Spence).[6] It is not so much a personal attachment to the past life as homesickness for the lost empire from its subject. Therefore, the book started from the Zhongshan Mountain and GBEP not to render the architecture but to yearn for the flourishing and powerful early days. In the constructed world of memory, the GBEP was transformed from an emperor's private monument to a symbol of a specific dynasty.

No one cared about the situation when Zhang Dai wrote these texts, they were accepted with delight. The reason is that Zhang's words satisfied people's imaginations and the need for historic monuments today; one of the most important dynasties in the history of Nanjing city – the Ming Dynasty – and its greatness, wealth and prosperity are still remembered today.

Therefore, from the beginning of the construction of the discourse, the GBEP was deprived of its original historical context, which had identified it as an imperial architectural piece in a great empire, a private monument of an emperor, a holy Buddhist relic and the imagination of missionaries. What was chosen and emphasized deliberately was a masterpiece of architecture that was compared with other architectural miracles in the world, and associated with a specific city and a specific period (Song and Yue). Although the architecture had been destroyed, the citation of ancient documents made the pagoda a hidden, historical marvel beyond itself oriented to a boarder space and time and entering the collective, public imagination. Thus, it became an object with a universal value.

However, the progress of the project was not quite smooth due to very limited investment and support from the NMG until 2008. At first, the project was expected to be carried out through private sector financing rather than government funding. However, an architectural miracle was not convincing enough for investors who made profit their first priority. Consequently, financing became the primary obstacle to progress. The estimated investment had increased to RMB 10 billion by the end of 2006 from RMB 1.3–1.5 billion in 2001. To push forward the project, the implementation entity – the government of the Qinhuai political district – founded the Jinlin Great BaoEn Temple Construction in 2003 (regardless of the RMB 1 billion investment), but there was still no improvement in the investment. Therefore, the crucial issue became how to attract the interest of the government and of private investors.

It was the archaeological discovery of a buried palace built in the tenth–twelfth century and the King Ashoka Pagoda in 2008 that pushed the GBEP to the focal point of the public voice all of a sudden and reframed its story. Moreover, the discovery of one of the most precious Buddhist relics in 2010 in the Pagoda ruins, the parietal Śarīra of Sakyamuni, proved the value of the site further. Thus, the history of the site, including the context of the Buddhist temple, the pagoda, the monks that once lived and the Buddhist relics, like the buried Śarīra of Xuanzang,[7] the Chinese Buddhist monk master's parietal, has been described carefully in order to emphasize the importance of Nanjing in the history of Buddhism (Zhu and Lu). So, it was no coincidence that, in 2010, a CEO invested RMB 10 billion in one go, which covered the expenses of the project's first phase. The archaeological discovery also led NMG to reconsider the scale of the project and it increased the budget to RMB 25 billion from RMB 10 billion.

A report published in the *Nanjing Daily* on 9 November 2010 was typical of the texts of the GBEP and other projects at the time.[8] In this paragraph, the discovery of early Buddhist remains and relics became the linkage between the pagoda and the city and its long history. It also broke through the limitations of a particular dynasty and established a layered history and memory of the city on site. From a single pagoda, its cohesive significance extended to the site, the city and finally the Buddhist world; the significance also extended to the history of the city since the sixth century from a single dynasty. Therefore, we can read from the texts that the history of the city was emphasized upon, portraying Nanjing as a significant city from the perspective of the south of the Yang-zi-Jiang River region, the country and the world. The pagoda was identified in the centre of this unfolding and extending space, and a long historical context. The original reason the emperor built the pagoda has never been mentioned. Thus, by manipulating a well-known government official concerned with the historical site, an empire and a city, an aura of significance was constructed around two identities – that of 'an architectural miracle' and 'a Buddhist holy land'.

The mass media, especially local newspapers, further fuelled this aura of significance and the progress of the project.[9] Media reports were also centred and organized around the project, and served as a platform to publicize the opinions of the government, scholars and citizens.

State-owned newspapers, like the *Nanjing Daily*, have a very special position. They closely track various government policies and decisions, and are an important channel of public access to city development news. Given their ease of purchase and reading access, citizens stay informed on various events of the city on time. This makes them feel like they are important participants and supervisors instead of just passive receivers in the public sector. It was the announcement of the reconstruction project by the *Nanjing Daily*, which got support from most citizens.

What is of similar importance is that the media focused on the significance of the pagoda, while the historical facts were treated as supplementary specifications under a 'relevant information' section in the corner of the related articles. So long as the details were provided, the significance of the pagoda would be self-evident. However, the historical 'facts' in the mass media were changed with the articulated significance of the pagoda. For instance, if 'an architectural miracle' was the focus, the historical 'facts' would be about the pagoda and the emperor, the height, the colourful architecture and a detailed description of 'one of the seven architectural wonders in the world'. When Nanjing as a capital of Buddhism was the focus, the stories of the King Ashoka Pagoda, the history of the different temples on the site, the buried Xuanzang monk master's parietal relics, and similar cases – like the find of relics claimed to be directly related to Buddha at Fa-Men Pagoda in Xi'an – were played up. In other words, the historical knowledge that people accepted from the mass media was organized and centred on select factors that could support the purported significance.

For most ordinary people, the GBEP is mysterious. Legend or rumours about the temple and pagoda emerged as early as before the end of the Ming Dynasty. For example, the first rumour was about the real identity of Emperor Yong-Le's biological mother.[10] Such gossip about the emperor's family pandered to the ordinary people's suspicion of official historical records.

Generally speaking, scholars, citizens, the media and various other players were both the creator and the receiver of the significance. First, architectural historians portrayed the GBEP as an important example of traditional Chinese architecture. After the initiation of the project, scholars constructed a richer and more complex image to correspond to the needs of the project. Also, the mass media not only satisfied the curiosity of ordinary people, it also pieced together obscure legends, fragments of ruins, place names and stories into a complete and meaningful narrative. More importantly, it also sifted through presentations of select narratives, and constructed and disseminated the meaning of the pagoda and the project, and organized relevant historical facts. The mass media thus guided or, more precisely, manipulated the public image of the pagoda.

Based on historical records, archaeological excavation and local folklore, various forces constructed the public image and discourses of the GBEP. The pagoda was transformed from a private monument to an eternal monument of the city. It not only became a symbol of a prosperous age, but also embodied the history of a city and a past empire. It is a part of the collective memory of the glory of the city, which makes the hidden historical monument obtain significance in present-day Nanjing.

## URBAN STRATEGY

### Substance and Discourse – Past and Present

Furthermore, the image of the GBEP created by different forces worked as an important means to establish the identity of the city today, which helped rationalize the reconstruction project.

The article 'Treasure the History, Re-create the Glory: Reviews on Architectures in Nanjing' published in *Study on Modern Cities* in 1995 (Pan 4–10) expressed some important, although unoriginal, opinions:

1. *Architecture and the times.* 'Architecture symbolizes the civilization of an era [ … ] represents the achievement of material civilization and spiritual civilization in different times' (Pan 4). As we all know, the opinion can be traced back to Hegel. As Ernst Gombrich has said, although we never talk about the self-consciousness of the spirit, we still regard artworks as an expression of the *Zeitgeist*. That allows an individual artwork with some kind of collective consciousness (Gombrich 3–9). According to this understanding, individual architecture must, at the same time, share common philosophical, political, social or economic conditions – the so-called spirit of the times. Thus, it endows an individual architecture with the ability to symbolize or represent a particular time or group.

2. *Architecture and the city.* 'Architectures are the body and appearance of a city. A city's history can be traced by its architectures built in different times, and its unique style and characteristic can be performed by these architectures' (Pan 4–5). That is to say, the visual presentation of a city depends on the material presence of architectures. The coexistence of architectures built over different times is the physical evidence of the history the city went through. A city possesses a distinct characteristic; it is not only a collection of architectures from different ages, but also a form of social community. The characteristic is manifested by the architectures and the collective consciousness of a city.

These views associate individual architecture with a time, a society and an urban community at the symbolic and physical levels as well as with the time stream of the past, the present and the future. From this point of view, the physical coexistence and visual presentation of architectures from different times transformed the collective history and memory of a city from abstract discourses to perceptible, understandable and communicable objects, thereby allowing the construction of a visible history of the city and establishing its unique characteristics. It seems these opinions have become the theoretical basis for the preservation and representation of historical monuments in Chinese cities today.

At the end of the article, the author not only described his expectations for future architectures in Nanjing, but also advocated the preservation of architectural heritage. He also hoped to 'restore some important historical architectures, such as the Ming Xiao Ling (mausoleum of the first emperor of Ming Dynasty), the GBEP and the Zhonghuamen Gate, for the public to visit and pay respect to' (Pan 10).

Similar to the group of important historical relics mentioned in this article, the reconstruction project of GBEP was not the only case related to the preservation, restoration or reconstruction of monuments in Nanjing. In the 'Master-plan of Nanjing City' issued in 2001, it was one of the three intentions of Nanjing to be 'a historical city with international influence and multiple culture characteristics'. As a part of the master plan, a new version of the *Preservation Plan of Historic City of Nanjing* was also published, which comprehensively sorted out historical heritage sites above ground, underground and in documents in Nanjing. Then the governments of Nanjing's political districts, especially those in the old city, implemented the strategy mentioned in the master plan through construction projects related to historical monuments. Most of these projects are like the GBEP and involve the preservation, rehabilitation and reconstruction of historical ruins. Like the GBEP, the value and significance of these projects have been carefully elaborated on in the representation process.

However, compared with the more than 500 historical heritage sites listed in the *Preservation Plan of Historic City of Nanjing*, the number of historical monuments repaired and reconstructed is still limited. Which of the listed monuments needs to be prioritized or is worth reconstruction or display? Some projects that were implemented at the same time as the GBEP include the Archaeological Site Park of Nanjing Homo Erectus Fossil, the Archaeological Park of Stone City, the Archaeological Park of the Ming Dynasty Palace and the Park of Niushou Mountain. Among them, the Archaeological Site Park of Nanjing Homo Erectus Fossil dates the history of Nanjing to the New Stone Age, and the Archaeological Park of Stone City and the Archaeological Park of the Ming Dynasty Palace are both royal projects from two important time periods in the city's history. The Park of Niushou Mountain is regarded as a significant Buddhist site. Other complete or nearly complete projects involving historical monuments, such as the former residence of Ganxi, the Yuejiang Tower and the Museum of the Jiangning Department of Imperial Fabric Weaving Supply, are all related with famous personalities of Nanjing's history. The value of a historical monument depends on its contribution to the construction of the city's history.

Ten years after the master plan was issued, the *Nanjing Daily* concluded that the 'modern construction of the city gave consideration to preserve the historical context and sort out cultural heritages, [and] highlighted the feature of [the] integrated modern civilization and [the] traditional culture essence of the city' (Xing and Zhu) during that decade.

Obviously, 'culture' and 'feature of the city' formed the core of the urban strategy. What this strategy was really concerned about was not the past, but the present. It aimed to construct a unique image of Nanjing different from that of other cities in that moment to 'improve the popularity' and then 'enhance the competitiveness' of the city. This image, as mentioned in an article published in 2009, should be 'made' (Ning, Chen and Zou). Ruins offered the opportunity for such 'making', production that contributes to shaping the features of the city would be carried out after preservation, restoration or reconstruction. According to the strategy, the GBEP project was one of the 'vitally important projects which would promote the influence and competitiveness of Nanjing in the world' (Li and Mao).

As a result, the Great BaoEn Temple and other similar historical monuments were dug out for a specific purpose whether at a discourse or a physical level. Thus, the writings about historical monuments by scholars and the mass media established connections between discourse and substance, the past and the present, and architectures and the city rather than establishing history and the meaning of an architecture or a city. Such connections rationalized the preservation or reconstruction of historical monuments in a modern city and, more importantly, made them an indispensable part of urban strategy.

## HISTORICAL MONUMENTS AS CITY ASSETS

Furthermore, the attitude of the government and the role that the GBEP and other similar projects played in the urban strategy should be more carefully examined in a broader background.

The reforms in the real estate and tax allocation systems, and market liberalization in the 1990s made local governments in China become economic entities with independent interests. Like an enterprise, a city has to deal with market competition to surpass peers (Zhao 7–15). Since 2000, 'managing a city' suddenly became a hot topic in academic discussions, policymaking and the mass media. In this context, competition among Chinese cities became increasingly fierce. In order to improve the competitiveness of a city, and attract investment and population, municipal governors needed to integrate and rationally allocate resources so as to maximize the gross urban economy.[11] In the past decades in China, gross domestic product (GDP) was the most important criterion to evaluate the achievements of a governor's career. It even resulted in competition between sub-governments. Given the dominant position of land and infrastructure in government assets, urban planning and development became the most important strategy tools for municipal governors in China to manage a city. Consequently, the past 10 years have seen an upsurge in the development of Chinese cities. Obviously, municipal governors in Nanjing were aware of this (Wang).

It is of equal importance that the historical monuments restored or reconstructed bring in considerable income for the city. Therefore, from the very beginning, the economic profit of the GBEP project was carefully calculated simultaneously with the building of its historical significance. In the proposal submitted to the municipal government, this project was targeted to be developed as a tourism site. Administrators believed that as a landmark and architecture of international repute, the GBEP had revenue-generating potential and would contribute to the economy of the city. Moreover, the archaeological discovery after 2008 further enhanced the expectations of the influence and economic benefit of the project.

The 'Jinlin Great Bao-En Temple Reconstruction Project Investment Invitation' issued in 2008 was an interesting document to this effect. It clearly showed that the GBEP project had multiple identities to the initiator. In this document, the historical significance – mixed with various views – of the pagoda and the role the reconstruction project played at the time – a complexity of religion, culture, history, tourism and commerce – were combined into one text. It seemed the executor of the project tried to establish a scenario of a cultivated modern lifestyle. However, in the last part of the document, the investment return estimates implied the real intention and operation mode of the project. According to the document, investment returns were expected to be generated from supporting the commercial space and land plot transactions. Thus, the essence of this project was tourism-related commercial real estate with a historical significance to develop commercial services, increase the land value and make profits from commercial operations or land transactions. Such tourism and real estate cross-industry projects were regarded as a profitable mode of development. From this point of view, all the significance regarding the time period, region, technology and religion were to transform the GBEP into a visualized commercial symbol so as to attract consumers (tourists), and improve the land value and commercial service quality.

Actually, benefit gambling among different forces ran through the project's progress and even resulted in a fight for the leadership of the project and the ownership of Buddhist relics.[12] Ten years after the project was initiated, with an expanding budget and a failure to attract private investment, the municipal government was even willing to get the initial capital through bank loans to kick off the project. The project manager implied that the GBEP and its archaeological site were recognized as city assets that were worthwhile for investment and could bring both social and economic benefits.

## WORK OF ARCHITECTS

The Great BaoEn project was assigned to a design team guided by architectural historians. The architects were expected to transform the abstract discourse to a physical and visible manifestation. The programme included the reconstruction of the pagoda, a new temple and a certain amount of commercial space. Hence, architects faced three issues in the design process: first, how to articulate the spatial relationship between the reconstructed pagoda and the archaeological site; second, how to connect the design proposal with the modern city; third, how to balance the different demands of a religious place, an archaeological site preservation and a commercial development (Figure 7.2).

Based on the events influencing the design strategy, the design process can be divided into three stages:[13]

1. From 2001 to 2008, before archaeologists confirmed the location of the pagoda and the extent of the temple, two concept proposals had been mooted. The main difference between them was the treatment of the new temple. In the second proposal, architects suggested that all possible underground relic locations – including the temple of the Ming Dynasty

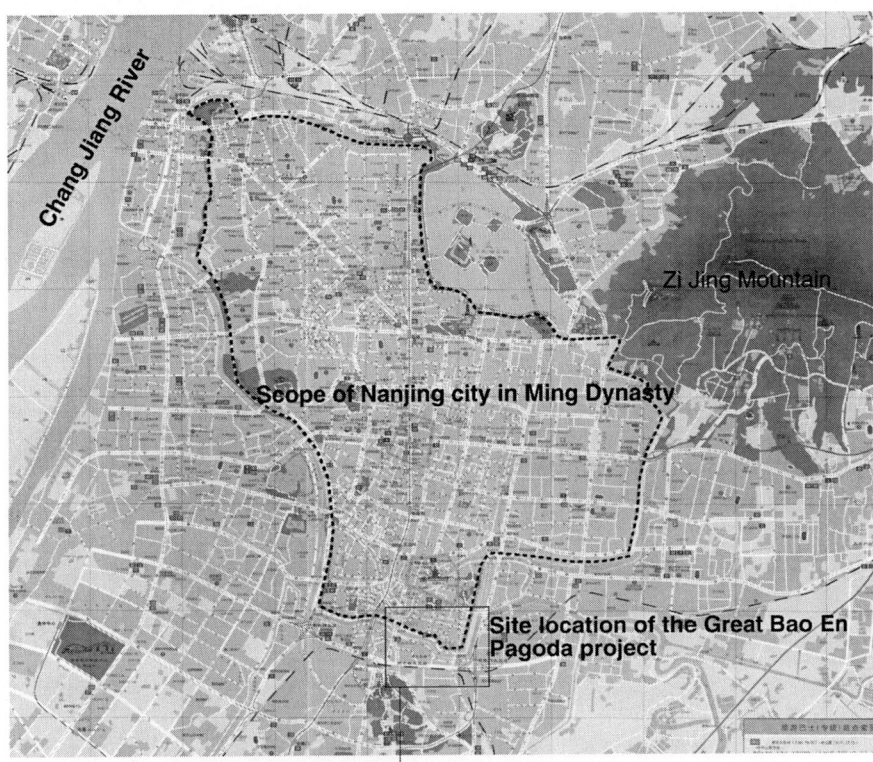

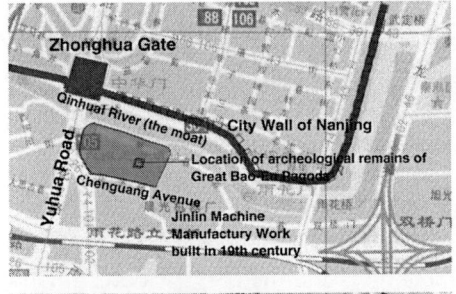

**Scope of the site in 2001**

In 2001 when the project started, the site area of this project was 7.6 hectares. The site is located in the south of the old city. North boundary is adjacent to the southern old city moat, west boundary is Yuhua arterial road extending out from the Zhonghua Gate, the southern city gate in Ming Dynasty, and one of the oldest development axis of Nanjing. South boundary is Chenguang Avenue and east boundary is the West Loop of Jinling Machine Manufacturing works, and its east site is the workshops of Jinling Machine Manufacturing works built in 1865.

**Scope of the site after 2008**

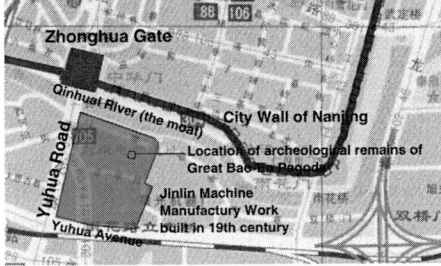

7.2   Location and scope of project site in 2001 and after 2008
*Source:* Jing Zhuge.

and the pagoda foundation – be reserved for exhibition purposes, while the new temple and the reconstructed pagoda line up along a new axis parallel to the speculated original axis from the Ming Dynasty. Since the archaeological park, the new temple and the new pagoda occupied the whole site, the commercial space was arranged under the new temple and the pagoda. The new temple was elevated a storey high on a platform that covered the commercial land in a huge indoor space. The other plan proposed a new temple – including a main hall and an enceinte – on the site where possible underground relics were located. The commercial space was located on the northwest side of the site.

Public reaction showed that there was a conflict between the public and the architects about how to deal with the ruins. From the point of view of the architects and architectural historians, the ruins of the old temple represented true history. They expected the public to build the true image of the GBEP by visiting the archaeological site. So they preferred the second proposal, which was better for preservation; also, by hiding secular commerce, the holiness of the Buddhist temple stood out. However, without professional knowledge, the ruins lacking strong, expressive forces seemed unappealing to the general public to whom visual presentation seemed more important than authenticity.

2. The archaeological discoveries of 2008 and 2010 changed the whole design strategy. On the one hand, the demand of the client – the execution body had at the time changed from the Qinhuai political district government to NSOAIM,[14] which meant the city administrator had become the consignor and the decision maker of the project – had changed. It asked that the design be centred on the shrine of the Buddhist relic of Shakyamuni's parietal and that the new design represent '[the] culture of [the] Ming Dynasty, [the] culture of Buddhism and [the] culture of gratefulness (the literal meaning of BaoEn)' (Li; Li, Wei, Gu and Wang). The site's boundaries were extended by two square kilometres. On the other hand, the discovery of temple ruins and the preservation demand made architects return to the design idea of the second proposal, which centred around the preservation and exhibition of the ruins.[15] Under the proposal announced in 2011, architects, archaeologists and preservationists worked together to construct an image of the pagoda which, together with the discourses, accomplished the construction of the city's history presented today. This was achieved by restoring the site authenticity and carefully designing the scenes, space sequence and spatial experience. It seemed that the authenticity the architects and preservationists pursued was finally realized and balanced with the visual representation that the public sought.

However, this time the area of the archaeological site was larger than expected, covering almost two-thirds of the original site. So the site was expanded southward. Half of the extended area was zoned for commercial purposes to compensate for the loss of the originally proposed commercial space now occupied by the park. The other half was expected to be developed into a cultural industry park.

3. After the decision to remove the Buddhist relic to another temple, the project plan was revised fundamentally. Instead of a newly-constructed pagoda, a 'light protective building in [the] form of a pagoda' (Zhu 2012), covering the archaeological ruins of the pagoda, became the only visual dominating the project site. The corridor in the traditional style half covering the temple ruins was transformed into a modern museum with contemporary construction material and technology. This final executive proposal displayed the layers and coexistence of modern construction and historical remains more clearly than the old ones.

Of course, as the critical part of the project, the reconstructed pagoda was always the focus of the design. However, the design of the new pagoda, which started with the restoration of an ancient architectural masterpiece, in the progress of the project and the design revision ended up as a skyscraper – an urban spectacle replacing the representation of a historical monument with modern technology and dazzling material. However, this also gave the new pagoda a clearer symbolic significance between the past and the present.

In the initial proposal, the new colour-glazed pagoda was expected to be an exact restoration of the original form and craftsmanship. However, problems cropped up soon. The 30-storey pagoda was as high as a skyscraper and elevators needed to be installed to transport visitors to the top. The newly added traffic core tube increased the area and size of every floor. Reporters said the pagoda was getting 'fatter'. In the following revision, the restoration of the original pagoda was actually given up and modern, coloured glaze production replaced the restoration of the original craftsmanship. The form of the pagoda pursued the features in the Ming Dynasty, a coordinated relationship in the height with the modern city but not a copy of the original. Meanwhile, in order to obtain a complete central space, the elevator placement was similar to the partial core barrel set in an octagonal flat side. The structure and space pattern of the new pagoda had no fundamental difference with modern skyscrapers like KPF's MidCity Place. Inside the new pagoda and temple halls, ostentatious luxurious adornments illustrate the Buddhist world. The architects, thus, changed their design strategy from copying the original pagoda to constructing a brand new skyscraper with features from the architectures of the Ming Dynasty – based on the knowledge of that dynasty and the technology of modern times – and created a colourful atmosphere of the Buddha with modern, electronic sound and light technologies.

Although the fighting over the ownership of the Buddhist relic later and the final decision of moving it to Niushou Mountain resulted in the cancellation of the new pagoda, the idea of a 'light protective building in the form of a pagoda' (Zhu 2012) on the archaeological remains of the old one went further in getting rid of the shackles of the historical image and committed to a gorgeous display of modern construction technology. The architects' explanation is that since a new pagoda in the past dynasties never copied previous ones on the site, the current new one needed a new design and identity that revolutionized the original intention of the reconstruction project. The new GBEP was no longer a reconstruction of a historical monument but an emergence of a new landmark in a modern city.

In this context, the GBEP in Nanjing was not the only such case in modern Chinese cities. It is also obvious that the GBEP project's consignor and architects were familiar with the precedents in other cities, such as the Leifeng Pagoda in Hangzhou, the Fan'gong Buddhist Temple in Lingshan Mountain, Wuxi and the Daming Palace National Archaeological Park in Xi'an. The Leifeng Pagoda was a reconstruction project of a famous ancient pagoda too. Completed in 2004, it was very similar to the GBEP project. After a national design competition, the Hangzhou municipal government chose a proposal to rebuild a new pagoda as a protective building on the archaeological site. Modern facilities like elevators were installed and the interior decoration was exquisite and splendid, but the exterior followed a typical medieval, Chinese pagoda structure. The vast Daming Palace National Archaeological Park in Xi'an is a display and protection project for a famous palace built in the seventh century. This project was highly praised and has become a model for many other Chinese cities. As for the splendid and magnificent Fan'gong Buddhist Temple, ancient Chinese and European traditional styles were mixed in the interior design creating such a luxurious Buddhist world that even Prince Siddhartha would have been stunned. The final proposal of this project was selected by tourists through a design competition. It is as great a touristic and commercial success as the Leifeng Pagoda project. These projects are already well-known in China, becoming hot attractions in their respective cities. In the revised plan for the GBEP, the ancient-style pagoda with modern materials and construction technology, and the imaginative interior design – coupled with the premise of authenticity, and the protection and display approaches of the archaeological remains – can be said to have a commonality with those precedents. Just like them, Nanjing's new pagoda and archaeological park are expected to become the new 'urban spectacle' of the city.

## CONCLUSION

First, after an assessment of the GBEP project in Nanjing, we can understand why the reconstruction of historical monuments became so popular during the past decade and a critical part of the development strategy of 'city management' in China that, behind the historical and cultural logic, it was driven by capitalistic considerations which were served by writing, advertising and designing. It is expected to increase city revenue and contribute to the political prestige of the city managers.

Against the background of the intense political and economic competition among cities, the construction of a collective history and memory of a city could give it a unique identity, thus endowing the historical monuments with value and meaning today. The visual representation transforms the monuments to new 'urban spectacles' and establishes a specific image for the city. In the fierce competition among modern cities, this uniqueness acts as a commercial branding that can effectively enhance the city's reputation, draw the attention of investors and attract capital and talent into the city. Therefore, they are regarded as assets – like land or infrastructure – that can generate actual profits. Furthermore, although there are various historical or cultural heritages in a city, only those shaping the city's history and meeting the capital demands of the operation will be chosen.

In most Chinese cities, these assets were or are being excavated physically and symbolically. Writing, advertising and designing these assets are necessary to transform them into products that can generate both political and economic benefits. In this context, the reconstruction or representation of historical monuments has become one of the most popular strategies in city management for China's contemporary urban managers.

Second, an assessment of the GBEP project also reveals that the mass media selectively sifted through discourses and built a false image of uniformity. When the urban strategy was implemented, various discourses disseminated through the mass media changed the meanings and magnified the significance. However, as mentioned earlier, the mass media chose history according to its requirements. It presented a uniformity of interest of different forces, such as architects, citizens, historians, investors and the government. However, the changing demands and situation along with the project process exposed its falsity. Moreover, the mass media never mentioned the fate of the people who lived on the site before the project was initiated. In other words, a lack of public critical space was implied from such false uniformity.

Last but not least, the role architects play in such projects is delicate. On the one hand, they are indispensable to such urban strategies because physical presentation is a prerequisite for transforming discourse into assets. On the other hand, architects' dependence on such a system of capital and political power makes them incapable of criticizing the strategies. The only thing left is to implement the plans with limited operational freedom and technology. Thus, we can understand the efforts of architects to balance the varying demands in the Great BaoEn project. Finally, architects – sometimes along with archaeologists and heritage preservationists – integrate the truth with imagination to carve a unique image and history of a city, construct its past, and also convert, by using expressive modern technology, represented historical monuments into new 'urban spectacles' with a cultural identity that gives a city a new collective memory.

## BIBLIOGRAPHY

Bernard, William D. *Narrative of the Voyages and Services of the Nemesis from 1840 to 1843: And of the Combined Naval and Military Operations in China*. London: Henry Colburn, 1844. Print.

Blussé, Leonard and Zhuang Guo-Dong. *Study on an Embassy from the East-India Company of the United Province to the Grand Tartar Cham Emperor of China*. Xiamen: University of Xiamen Press, 1989. Print.

Cai Yan. 'The Great Bao-En Pagoda Will Be Represented in Jinling in 3 Years'. *Jiangnan Times* 23 Sep. 2001. Print.

———. 'The Great Bao-En Pagoda Will Be Represented in Nanjing'. *Lanzhou Morning Post* 14 Oct. 2001. Web. 25 Nov. 2011 <http://lzcb.gansudaily.com.cn/system/2001/10/14/000325903.shtml>.

Chen Yan-Fei, Chen Si-Ping and Lu Yong-Mei. 'The Great Bao-En Pagoda Will Be Represented in the Old City'. *Nanjing Daily* 20 Sep. 2001: Sec.3. Print.

Gombrich, Ernst. 'Hegel and Art History'. *Architectural Design* 51.6–7 (1981): 3–9. Print.

He Qiaoyuan. *Ming Shan Cang*. Fuzhou: Fu Jian People, 2010. Print.

'Jinlin Great Bao-En Temple Reconstruction Project Investment Invitation'. Information Office of Nanjing Municipal Government, 2008. Web. 19 Nov. 2011 <http://gov.longhoo.net/2010–09/09/content_4059642_2.htm>.

Le Comte, Louis. *Nouveaux Mémoires sur l'état Présent de la Chine*. Paris: Jean Anisson, 1696. Print.

Li Ji and Mao Qing. 'Making a Series of Signature, Cultural Production Projects High Level'. *Nanjing Daily* 6 Nov. 2010: Sec. A01. Print.

Li Ji, Wei Ming, Gu Ping and Wang Jian-Lin. 'Contributing RMB 10 Billion for Traditional Culture'. *Nanjing Daily* 9 Nov. 2010: Sec. A02. Print.

'Master-plan of Nanjing City'. 2001. Web. 16 Oct. 2013 <http://www.njghj.gov.cn/ngweb/Page/Detail.aspx??CategoryID=dd8b3d29–1b2a-4d65–8275–496d16e6e7b6&InfoGuid=228fcddd-825c-4287–8ba9–7781953a23a5>.

Nieuhof, Johannes; Pieter de Goyer; Jacob de Keizer; John Ogilby; Johan Adam Schall von Bell and Athanasius Kircher. *An Embassy from the East-India Company of the United Provinces, to the Grand Tartar Cham, Emperour of China*. London: John Macock, 1669. Print.

Ning Jian-Xin, Chen Yan-Qiu and Zou Wei. 'Preserving the Old City, Creating the City's Characteristic'. Urban Construction, *Nanjing Daily* 29 Jul. 2009: Sec. A04. Print.

Pan Gu-Xi. 'Treasure the History, Re-create the Glory: Reviews on Architectures in Nanjing'. *Study on Modern Cities* April (1995): 4–10. Print.

Qiu Baoxing. 'Managing City and Competitive Power of City', *Journal of Party School of CPC* Nov. (2001): 84–8.

Semedo, Alvaro. *The History of That Great and Renowned Monarchy of China*. London: John Crook, 1655. Print.

Song An and Jie Yue. 'Pay Attention to the Great Bao-En Pagoda'. Special Reports, *Nanjing Daily* 22 Sep. 2001: Sec. 7. Print.

Spence, Jonathan D. *Return to Dragon Mountain of a Late Ming Man*. New York: Penguin Books, 2008. Print.

Survey Group of Theory Frontier of Party School of CPC. 'Developing the City, Managing the City, Shaping the City', *Theory Frontier* 22 (2000): 21–4. Print.

Wang Run-Juan. 'Innovating Thinking, Progressing Overall, Starting Comprehensive'. *Nanjing Daily* 5 Dec. 2001: Sec. 12. Print.

Wright, George N. and Thomas Allom. *China, In a Series of Views: Displaying the Scenery, Architecture, and Social Habits of That Ancient Empire*. London: Fisher, 1843. Print.

Xing Hong and Zhu Kai. 'Fruitful Construction of Cultural Nanjing'. *Nanjing Daily* 8 Jan. 2011: Sec. A07. Print.

Zhang Dai. *Memory of Tao An*. Peking: Zhonghua Book Company, 2007. Print.

Zhang Hui-Yi. *Records of Jinlin Great Bao-En Pagoda*. Nanjing: Nanjing Press, 2007. Print.

Zhao Yan-Jing. 'From Administrate a City to Manage a City'. *Urban Planning* 26 Nov. (2002): 7–15. Print.

Zhou Qingshu. 'Notion of Destination from the View that Hongjila as Bilogical Mother of Emperor Chengzu'. *Journal of Inner Mongolia University (Philosophy and Social Science)* Mar. (1987):1–28. Print.

Zhu Kai. 'The Site of Great BaoEn Temple was Listed as one of the 10 Major Archaeological Discoveries in China'. *Nanjing Daily*. 10 June 2011: Sec. A01. Print.

———. 'Answers to Reporters' Requests by the Person in Charge of the Great BaoEn Temple and Niushoushan Mountain Archaeological Site Park Projects'. *Nanjing Daily* 16 Sep. 2012: Sec. A01. Print.

Zhu Kai and Lu Yong-Mei. 'Ashoka King Pagoda Will Bring about Development for Nanjing as a Centre in the Buddhist Field'. News of Entertainment, *Nanjing Daily* 27 Nov. 2008: Sec. B07. Print.

## NOTES

1   A systematic study on the Great BaoEn Pagoda began in the 1980s. Architectural historians studied and restored the form of the pagoda. They praised its construction technology and coloured glaze craftsmanship highly, and called the coloured glazed pagoda an important example of traditional Chinese architecture. Interestingly, those historians were architects too. They not only supported the reconstruction of the pagoda, but also guided the design in the reconstruction project. This pagoda was called the Porcelain Tower. So, the other main topic that scholars favoured was the influence of the Porcelain Tower on Europe. In the study on the influence of Chinese art on European gardens and architecture, the Porcelain Tower was one of the acknowledged models. However, writings on the Great BaoEn Pagoda in travel books by European authors were researched only after the reconstruction project started. The city's importance in Buddhist history did not get much attention until archaeological excavation began in 2007. Then, the discovery of an underground palace built in medieval China, King Ashoka's pagoda and the Buddhist Sakyamuni's parietal relic encouraged many articles after 2000.

2   See *The History of That Great and Renowned Monarchy of China*, written by Portuguese missionary Alvaro Semedo (1585–1658), *An Embassy from the East-India Company of the United Province to the Grand Tartar Cham Emperor of China* (Dutch and French versions published in 1665), written by Johan Nieuhof (1618–72), a member of the Dutch diplomatic corps, and *New Memoirs on the Present State of China* (Paris, 1696), written by French missionary Louis le Comte (1655–1728). These books were written on the basis of the authors' personal experience, and had a great impact on the Sinomania in Europe in the eighteenth century. All the authors have been to Nanjing. However, Alvaro Semedo probably did not visit the Great BaoEn Temple. He was not only wrong about the number of storeys of the pagoda, but also sounded unsure in the narrative. However, he said that the architecture could be regarded as one of the most famous architectures in Rome. Nieuhof and Louis le Comte had, obviously, visited the temple and climbed up to the top floor of the pagoda. Nieuhof's work was regarded as the earliest and most valuable resource of Chinese knowledge in the seventeenth century.

3    The most detailed description of the pagoda came from Louis le Comte, and Johan Nieuhof described the view from the top floor of the pagoda. In *China, In a Series of Views, Displaying the Scenery, Architecture, and Social Habits, of that Ancient Empire* (co-edited by George Newenham Wright and Thomas Allom, published in 1843), the editors quoted texts from Alvaro Semedo and le Comte. The pictures selected by the editors were of the pagoda, the temple's main hall and the view from atop the pagoda. The record in *Narrative of the Voyages and Services of the Nemesis from 1840 to 1843, and of the Combined Naval and Military Operations in China,* written by W. D. Bernard in 1844, was the last time the Great BaoEn Pagoda appeared in European writings before its destruction. In this book, the beautiful colour of the glaze and the magnificent view from the top of the pagoda were still the main attractions for the British officer.

4    Actually, in the original text, Henry Nieuhof said the pagoda was parallel to the Seven Wonders of the World, and not one of them.

5    This book was written after 1644 and first published in 1794.

6    After the Ming Dynasty fell, Zhang Dai travelled over the mountains in the southwest of Shaoxing. His difficult circumstances were a marked contrast to his luxurious life in the past. So, he tried to recall the old empire and his old life through his writings.

7    Xuanzang was a famous Buddhist monk in the seventh century in China. He took 19 years to travel to India in order to study Buddhist scriptures and brought back many important texts to China.

8    Nanjing was listed as a historical Chinese city by the Ministry of Housing and Urban Rural Development of the People's Republic of China. There is considerable historical and cultural heritage in this city. The Jinlin Great BaoEn Temple was located in Chang-gan Li outside the Zhong-Hua Gate. In the past 18 centuries, Nanjing has remained a centre of Chinese Buddhism and the birthplace of the Buddhist temple in the south of the Yang-zi-Jiang River region. In 1412, the 10th year of the Yong-Le Empire, the emperor decided to rebuild the Jinlin Great BaoEn Temple and a nine-storey pagoda in coloured glaze. It took 19 years before Zhen He inaugurated the new temple. In the following four centuries, the Great BaoEn Pagoda was regarded as the most distinctive landmark in Nanjing and was called 'the No.1 Pagoda in the world' (Li, Wei, Gu and Wang). It was named the 'Porcelain Pagoda of Nanjing' by the Western world because of its lustrous and colourful glaze. It was one of the seven miracles of the medieval world, together with architectural masterpieces such as the Colosseum in Rome, the ancient city of Alexander and the Leaning Tower of Pisa, and was copied in Western countries as Chinese classical architecture. In 1856, the Great BaoEn Temple was destroyed in the Peace Heaven War. On 12 July 2010, the only remaining Buddhist relic of Shakya Muni's parietal was found 1,000 years after it had vanished. In accordance with the expectations of the general public, the municipal government of Nanjing decided to rebuild the Great BaoEn Temple and its coloured glazed pagoda to represent the spectacle of a Buddhist centre and to preserve the 1,000-year-old history and culture of the city. The Great BaoEn Pagoda Reconstruction Project has been listed as one of the 16 key projects in Nanjing in 2010 (Li, Wei, Gu and Wang).

9    For example, it was the public report and propaganda that made the archaeological excavation of the Great BaoEn Pagoda a much-discussed topic for the general public.

10   Zhou Qing-Shu pointed out that this legend came from the *Records of Taichangsi in Nanjing* written by Wang Zong-Yuan in 1623. However, it was published first in *Ming Shan Cang* written by He Qiao-Yuan after 1596. He Qiao-Yuan cited the story from the *Records of Taichangsi in Nanjing*. After that, many scholars quoted it. Consequently, more and more people believe the story.

11   In 2000 and 2001, two articles published in important national newspapers or journals indicated a turning point in the national city development policy. These articles discussed the topic of 'managing a city' or 'running a city like an enterprise'. The patterns they summarized from some earlier cases have become a general development pattern for 'managing cities' in China since 2000. It includes some important aspects: investment invitation for infrastructure construction, setting up high-tech development zones as new economic growth points of the city, innovation of property rights of state-owned enterprises and optimizing urban-centred organizations of regional economies. Survey Group of Theory Frontier of Party School of CPC; Qiu.

12   In 2001, the political district governments of Qinhuai and Yuhua submitted their applications for the Great BaoEn reconstruction project almost simultaneously. In 2010, Master Chuanzhen, a Chinese People's Political Consultative Conference (CPPCC) member in Nanjing and the abbot of the Xuanzang Temple, suggested that the Great BaoEn Temple be rebuilt in Niushou Mountain (another important Buddhist holy land in Nanjing). Finally, the municipal government decided that the Buddhist relic, the parietal bone of Sakyamuni, should be enshrined and worshipped in Niushou Mountain. This final decision derived its religious significance from the Great BaoEn Temple, but will bring great touristic reputation and appeal to the less well-known Niushou Mountain Park. Thus, this decision will increase the total income of the city. It was clearly not a coincidence that the Great BaoEn Pagoda project and Niushou Mountain Park held their groundbreaking ceremonies on the same day.

13   Render drawings and the master plan of the project proposals of 2004 and 2008 were posted on the website of the Nanjing Urban Planning Bureau. Revised proposals after 2010 can be found in the unpublished yearbooks of the Urban Architectural Lab (UAL) studio.

14   Nanjing State-Owned Assets Investment Management Holdings (Group) Co. Ltd. This is a company founded by the municipal government of Nanjing.

15   In addition to the famous underground palace built in the tenth and the twelfth centuries and the Buddhist relic, the archaeological excavation also revealed the layout of the Great BaoEn Temple. In fact, the discovery of the temple and the underground palace was one of the main reasons that the archaeological excavation of the Great BaoEn Temple was listed as one of the '10 major archaeological discoveries in China' in 2010 (Zhu 2011). This resulted in intervention from the State Administration of Cultural Heritage. The archaeological site was included under the legal network of cultural heritage preservation. Therefore, the preservation and display of the archaeological site became an unquestionable decision.

# 8

# Architecture, Reconstruction, Memory: The Image of Tel Aviv's White City

*Alexandra Klei*

For several years now, Tel Aviv has been inseparably associated with the terms 'White City', 'Bauhaus city' and hence with modernist architecture in general (see, for example Rössler). The reasons for this lie in an architectural heritage dating from the 1930s and 40s, comprised of roughly 4,000 structures. This heritage is also defined by the founding of the city no more than two decades prior, in 1909, originally as the Ahuzat Bayit settlement north of the port city of Jaffa.

Over the course of the past 30 years, a number of actors have initiated various processes to embed a certain image of the White City in the public's consciousness, to promote a certain sensibility for its architecture and, in the process, to provide the city itself with a unique point of interest and contribution to its identity. In 2003, these efforts culminated in the city's recognition as a UNESCO World Heritage Site. A look at the city as it appears in reality, however, reveals that the architecture in question is in need of some reconstructive effort in order to conform to the construct of its own image after years of neglect and considerable redevelopment. Also, there are numerous clues to other layers and histories of urban development in evidence.

This chapter discusses the significance of both extant and reconstructed architectural traces and their influence on the way their history is conveyed to the general public. To start with, some deliberations on architecture as a medium of memory, the concept of reconstruction in terms of creating an architectural image and the role of architectural reconstructions in forming a cultural memory will serve as an introduction to the subject.

## ARCHITECTURE AND MEMORY

Architecture as a medium of memory commands both a direct and an indirect potential for remembering, which is to say that it is capable of triggering memories on the one hand but also of manifesting and storing them on the other

(Will 113–32). On the one hand, an observer might recall, in looking at a building, a personal experience related to this place or associate it with an historic event. This process can be reinforced with the aid of external media such as texts and images on information panels, guided tours, brochures, public mementos and so on – anything that adds emphasis and meaning to a building or a public space and makes it stand out from its surroundings. Such labels are preceded by a (re-)discovery of a place's history and are a result of subsequent processes initiated by actors that are intended to create public reminders and make information more visible for outsiders. This indicates that architecture serves to both localize and explain events. It is important to remember that this kind of meaning attached to architecture is a construct: what is conveyed about a place and its history is what certain actors intend to convey.

On the other hand, the building itself stores information, regardless of whether or not observers perceive and decode it. For example, information may be preserved about an historical period and its achievements in terms of creativity and engineering, about societal norms, needs and intentions – but also, through additions or redesign efforts for instance, about change and the relationship between the private and the public or, regarding entire neighbourhoods, about their size and boundaries with their surroundings. This, in turn, may lead to assumptions about how buildings and neighbourhoods have been used, about their inhabitants' social status and so on. All of this means that such assumptions can still be made based on the actual architecture itself, even in the absence of media explaining the origins of a certain building or part of a city – a fact that is especially relevant regarding the purpose of this chapter. Architecture preserves this kind of information and may act as an historical cue that triggers memories of the past or even lead to the subsequent rediscovery of a place.

## ARCHITECTURE, MEMORY AND RECONSTRUCTIONS

To reconstruct something means to reproduce, recreate an 'original state' (*ursprünglichen Zustand*) and to 'identify down to its details and accurately portray, depict' something that is past (in seinen Einzelheiten [zu] erschließen und genau wieder [zu]geben, dar[zu]stellen).[1] Consequently, this opens up the possibility, among other things, to make something comprehensible, to convey a concept and to gain some new understanding.

In architecture, 'reconstruction'[2] implies that some time must have elapsed between the destruction and the recreation of an architectural structure, which sets it apart from the concept of rebuilding (Fischer 7–15). As a basis for a reconstruction, scientific analyses of 'spoken, image and written sources' (*Wort-, Bild- und Schriftquellen*)[3] are used, which are period-accurate only in select aspects. Ursula Baus correctly observed that, with reconstruction projects, 'there is, in most cases by far, only a call for a certain outward appearance – not for an historically accurate spatial arrangement, historically accurate functionality or all of those at once in the context of a forensic research effort across time' (Baus 105).[4] More precisely, as evidenced by recent cases in Germany (for example Brunswick Palace, Berlin City Palace or Dresden's Neumarkt), the interest in an accurate 'outward appearance' is frequently limited to whichever part of a building faces a public space.

Architectural reconstructions are the result of a selection process – not every building is supposed to be re-erected/recreated, not every layer of history preserved – and they are a product of the construction of meaning with reference to historical events or by singling out a particularly notable monument of a given period. They have to fit into the narrative that establishes, according to certain actors,[5] the necessity of a reconstruction. Thus, reconstructions have to be viewed as manifest acts of remembering historical events or periods. In the case of Tel Aviv, this is what is evidently happening with 1930s and 40s architecture regarding the term 'White City' and the associated narrative of the founding of the first modern Jewish city.

Within the context of a transition between communicative and cultural memory,[6] architectural reconstructions can be categorized as the latter: much like with archives or memorials, select aspects of history are intended to be transferred into a more lasting form of preservation and thus be made accessible for generations to come. By analogy with the debates about the construction of memorials and monuments, the question has to be asked not only what architectural reconstructions are supposed to recall and in what way they do that, but also what they suppress.[7] Regarding the aspects of selection, the assigning of meaning and its expression in architecture, the concept of reconstruction is therefore applicable to Tel Aviv's White City. In the interest of gaining new insights in operative and discourse-analytical terms as well as into the relationship of architecture and memory, this concept can be used to inquire about what information about a time period, an event and a style of architecture is embedded and retained in such a way as to make recalling it possible. This allows us to include questions about the construction of images and concepts as well as their expression into concrete architectural space and that space's subsequent character as a medium as well as its legibility. These processes have to be understood as reconstructive measures including mental and virtual constructs. As literal *re-constructions*, these projects would thus aid us in understanding both the construction of certain images and of physical buildings and urban ensembles, and would not be limited to the question of how much time has passed since their destruction or the dependency on other sources. In this context, it makes sense to use the term 're-construction' in order to differentiate between the merely physical reconstruction of lost architecture and the mechanisms, strategies and processes involved in reconstructing architecture. This also allows us to elucidate constructive aspects – aspects of selection and construction – involved in those processes as well as the actual structures.

## TEL AVIV – CITY HISTORY

Many texts on the history of Tel Aviv[8] start on 11 April 1909. It was on this day that the Ahuzat Bayit housing association drew by lot 60 families who were to each receive one plot of land.[9] The settlement would go on to bear the association's name for a year before it was renamed *Tel Aviv* ('spring mound'), the title chosen for the translation into Hebrew by poet and writer Nahum Sokolow of Theodor Herzl's novel *The Old New Land* (1902). The association's members and initial builders were predominantly Russian immigrants who had until then lived in Jaffa. Planned as a suburb in the style of a European suburban garden city, the place was to provide a better standard of living than the crowded Arabic port city. At the new development project's centre was the first Hebrew grammar school, named after Theodor Herzl. Designed by Joseph Barsky, the building with oriental-style elements in its façades, which were intended to convey an 'old-new Hebrew character',[10] was also the most conspicuous structure there. Both the layout of the single-family homes and their gardens were unregulated, and so their construction was governed by their inhabitants' financial situations as well as aesthetic predilections. The new settlement grew rapidly, and the number of homes doubled within four years after its founding. Development plans played hardly any role.

The historical texts then move on to the period after the First World War, when the number of new inhabitants kept increasing and Tel Aviv became a city in its own right in 1921. Its appearance was now increasingly characterized by multi-family homes of architecturally eclectic design.[11] At that time, the city made two attempts to implement development plans. First, architect and city planner Richard Kaufmann, who had emigrated from Frankfurt in 1920, was commissioned. Even though he succeeded in devising a plan in no more than two months, it was never implemented, as rapid development was more in the city's interest than enforcing a long-term plan. In 1925, Scottish biologist and urban sociologist Sir Patrick Geddes was commissioned. Again, his plan to create a hierarchical system of streets involving large thoroughfares and calm residential streets as well as blocks of flats featuring semi-public facilities for a social infrastructure between neighbourhood units could not be fully implemented as intended but did serve as the foundation of the city's infrastructure still seen today. However, as Geddes's plans were based on a population of no more than 100,000, they were doomed to crumble in the face of the following waves of immigration: from a population of 60,000 in 1932, the count grew to 120,000 in the following three years. Among the immigrants were architects influenced by the new principles of modern European architecture. Given the chance to fully implement them here, they erected roughly 4,000 buildings in that style, mainly during the 1930s, but up until the early 50s. These had gleaming white to beige façades, deeply recessed balconies or loggias, accessible flat roofs, with some houses suspended on pillars atop green gardens. Their stairwells were emphasized by prominent vertical windows. Using these elements, they made reference to the five points – columns, roof garden, free ground plan design, free façade design, elongated windows – Le Corbusier defined in the 1920s as characteristic of his new architecture and adapted them for a Mediterranean region and its specific requirements. The balconies were supposed to provide shade for the inhabitants, the flat roofs as well as the shady entrance areas were meant to serve as semi-public spaces, and raising the main bodies of the buildings on stilts was intended to provide the city with better air circulation.[12]

It is these buildings that figure as the third part and conclusion in historical texts about the city, providing a basis for linking the city's history not only with its architecture but also with the White City concept.

## HISTORY PRESENT IN THE CITY

There is only a very small number of buildings left from the first phase of city development.[13] They are few and far between, in a bad condition and not clearly marked within public space. The importance of that phase in the city's history has yet to find an adequate expression in the way the remaining structures are handled. Also, the early settlement's former location and boundaries within the city are not evident. However, the Shalom Tower, an office tower opened in 1965, marks the spot where the grammar school made way for the tower's construction in 1962, and its presence commemorates the existence and significance of the original school building (Altrock 6).[14] Moreover, the western wing of the building hosts the Discover Tel-Aviv Center with a number of permanent exhibitions dealing with the city's architectural heritage, in which individual buildings, the Ahuzat Bayit settlement and a comprehensive view of Tel Aviv are represented by scale models, among other exhibits.[15] Furthermore, there are photographs on display of select specimens of modernist architecture in Tel Aviv as well as plans and very comprehensive material on the Shalom Tower itself.

Unlike the Ahuzat Bayit buildings, those from the eclectic phase have become an integral part of the way the cityscape is presented.[16] Even though some of them are still in dismal condition, others have been restored comprehensively and with a lot of attention to detail. Also, for a select number of them, information panels are provided, lending particular weight to their significance and integrating them into the narrative of the city's history. While the increased amount of attention spent on them in the past few years is clearly second to that dedicated to the White City buildings, there is a connection between the two: in the wake of the new discourse about the roughly 4,000 modernist buildings during the past 30 years, general awareness has been raised of an architectural heritage beyond sites dating from antiquity.[17] The main focus however is clearly on the buildings in the International Style and especially on those built in Tel Aviv and defining its cityscape.[18]

## FROM AN IMAGE TO A CITY: THE RECONSTRUCTION OF THE WHITE CITY

In 1984, on the occasion of the 75th anniversary of the founding of the city, the Tel Aviv Museum hosted the exhibition *White City: International Style Architecture in Israel*, curated by Michael Levin, which also covered Jerusalem and Haifa besides Tel Aviv.[19] However, in the following years, efforts to document the buildings – resulting in research into their architectural history – were initiated not for the former two cities, but only for Tel Aviv. Another exhibition on *Neues Bauen* between 1930 and 1939, this time in Germany by the Institut für Auslandsbeziehungen (Institute for

Foreign Relations) and the Munich Museum of Architecture, in the early 1990s was limited to Tel Aviv as well. Its catalogue, published in 1994, details – alongside 6 short essays and 66 biographical notes on architects – 83 buildings using black-and-white photographs of one façade in its then state, some notes and a (in some cases rough) ground plan. This publication indicates not only that considerable research into early architectural history was taking place, but also the prevalent interest German researchers took in the subject (Institut für Auslandsbeziehungen, Stuttgart, and Architekturmuseum der Technischen Universität, Munich), thus demonstrating the rapid internationalization of interest in the architectural heritage of Israeli modernism. The fact that the exhibition went on to be shown in 18 cities in Germany, Europe, the US and Israel, as well as in Turkey, in the following years only corroborates this.[20]

In Tel Aviv itself, the Bauhaus Center, founded in 2000, served to encourage discourse on the subject in a multitude of ways – exhibitions, events, guided tours of the city, books,and also the works of Italian designers – and at the same time supplied products that made it accessible for a lot of people. In addition, a Bauhaus museum has been welcoming visitors since 2008, focusing however more on design and displaying only a few historical black-and-white photographs.[21] These efforts found their culmination in the application for and the subsequent induction into the list of UNESCO World Heritage Sites of 'White City of Tel Aviv – the Modern Movement' in 2003, declaring three officially protected areas (A, B and C) categorized by time period (*White City of Tel Aviv: the Modern Movement*). This was accompanied by an exhibition originally presented in Tel Aviv but since brought to numerous cities worldwide.[22]

While the existence of an ensemble of 4,000 buildings is highlighted in all of these exhibitions, publications and so on, some examples have to be singled out, explained and assigned a special status in the interest of visualization.[23] This strategy helps to make the ensemble easier to conceptualize, while its constituent elements and history are verbalised and made more accessible to the observer. Select buildings tend to represent the whole, and at the same time differentiations and variety in the architectural solutions are lost.

All in all, the various media and initiatives primarily create an image, a concept of the city's architecture as white and modern, characterized by a clarity of form and proportion. They also emphasize a relationship with Europe, and by using the term 'Bauhaus' with Germany in particular[24] – a relationship with an architectural period with positive connotations. This links the city to a certain history and allows it to be categorized into a broader context of architectural history.[25] Taking into consideration the great number of buildings and the adaptation of architectural shapes and elements to the climatic characteristics of the region, they provide the city with something unique. The specific architecture becomes a vessel of an urban identity, which projects images and concepts to the outside. As the construction of images takes place in various media, these images also reach people far away who do not know the place first hand.

Often in these processes, historical pictures are used that are not quite indicative of the actual architecture, which is readily apparent when looking at the buildings themselves, many of which are still in dismal condition. This is due to the fact that they had hardly seen any repair in recent decades, that the basic structure of the buildings was in poor condition and numerous redesigns – very frequently the closing down of balconies or the addition of structures to the roof – had taken place and fundamentally changed their appearance. In this situation, it became necessary to reconcile the actual architecture with its perceived image.

Starting in the 1990s, but with the inclusion as a UNESCO World Heritage Site at the latest, efforts to reconstruct the built space were launched. The conservation plan adopted in 2008 identifies 1,000 historical buildings for which – in order to create an incentive for renovation and compensate for relinquishing the right to repurpose the real estate – permission was given to add storeys. For 180 buildings, no alterations are permitted (*White City of Tel Aviv*; Veser 2). The additional storeys are also intended to create new living space urgently needed in the city. In that regard, Tel Aviv is meant to be seen as a 'living city' adding another layer to its cityscape.[26]

That way, the concrete measures taken in the built space in the context of re-construction entail the extant structures' conservation and an increased appreciation of architectural heritage, leading to a clearer visibility of the White City. In terms of memory, the city for one thing makes information about its original appearance available again and also adds new information referencing the current inhabitants' needs. By making the buildings clearly distinguishable into old and new, change remains visible. At the same time, some information is bound to get irretrievably lost in future: adding storeys leads to a shift in proportions. While the original dimensions remain visible in individual buildings, the overall impression of an historical ensemble vanishes.

## MARGINALIZED CITY HISTORIES AND RECONSTRUCTIONS BEYOND THE WHITE CITY

An attentive observer walking the city's streets is bound to make out parts of the city from other than the architectural periods mentioned earlier and thus further aspects of the city's development, yet the publications available today reveal little about them. Sharon Rotbard critically examined the White City label in his 2005 book *White City, Black City* and deconstructed developments in the city and its self-perception as the White City in light of its political significance regarding Zionism and the Israeli/Jewish self-image. He also made mention of other settlements, especially Arab villages once located in today's urban area.[27] The Zochrot initiative has been working for years to make their history as well as the plight of their fugitive and even displaced Arab population and the destruction of their homes a part of the public perception of the city's history. They list four such villages[28] within today's immediate urban area of Tel Aviv.[29] The initiative's activities are aimed at the documentation and preservation of information in order to keep the memory of that history accessible, among others in virtual form. Also, there is still evidence of their existence in the cityscape itself. The most prominent example is al-Manshiyya, which originated in the late 1870s and grew into one of the largest neighbourhoods in Jaffa. In 1948, most inhabitants fled or were displaced, while some still lived there in the 1950s. The buildings were demolished starting in the 1960s (*Zochrot*). The area has been partially built over by office buildings and a shopping centre, with Charles Clore Park stretching along the Mediterranean coast. However, the overall impression is that the area's boundaries and interconnections with its surroundings as well as relationships of functionality and infrastructure are rather undefined. Two of the original buildings remain but are not conclusively integrated into the context of the city's history: Hassan Bek Mosque[30] and ruins partially built over in 1983 by the Etzel Museum (Cuéllar).[31] Even though they do not clearly delineate the full extent of the neighbourhood, they do hint at its dimensions, an impression reinforced by the fact that the park does not only allow for a clear line of sight between the two buildings but also indicates a gap that does not follow any clear spatial rationale. On the one hand, the current use of the ruined building as a museum about the history of the national Jewish military organization Irgun Tsvai Leumi ('Etzel' or 'Irgun' for short) indicates its integration into the Jewish narrative of the 1948 fighting, while on the other hand the remaining ruins are evidence of the prior presence and destruction of the Arab neighbourhood. The mosque is, in both function and symbolic character, an unmistakable and highly visible indicator of a (previous) Arab presence in the area and stands out from its surroundings. The way the area is now used attaches new meaning to the place but cannot consign its history to oblivion. It offers up a variety of interpretations and memories, which however require the perception and appropriation of actors taking an interest to be more clearly defined and categorized.[32]

Immediately adjacent to the east are further spaces calling to mind a complex history and processes of superposition and appropriation, for instance the recently restored area of the former settlement of the Hugo Wieland Templer family built after 1900 and the adjacent Jaffa–Jerusalem line train station opened in 1892 (Woelke). Not only have the area's 22 buildings been extensively renovated in the context of their rediscovery and put to new use as cultural and commercial infrastructure, they also have been provided with information panels detailing their origins as well as explaining their connection to the history of the area's early development. This part of the city does not only stand out due to its architecture – detached buildings, some with gabled roofs and shutters[33] – but also because of the dense historical narrative provided by the many panels.[34] This is very different from Neve Tzedek, the first Jewish settlement, dating from 1887.[35] Here, the area is clearly differentiated from its surroundings by its closely spaced, low houses separated by narrow streets, some winding – an established, renovated neighbourhood that travel guides say 'brims with galleries, museums, restaurants, cafés, bars and small shops' (*sprudelt voller Galerien, Museen, Restaurants, Cafés, Bars und kleinen Geschäften*) (Stein 80). The buildings' restoration was largely unregulated in terms of historical conservation, and as a consequence there is little visible distinction between original and new parts (see Altrock 26). Also, apart from its architectural appearance, there is hardly any reference to the area's history: only a few information panels exist, such as a billboard that has been directing attention to the planned renovation of Aharon Chelouche's former private residence, one of the founders of the settlement.

One travel magazine claims that Ahuzat Bayit merged with two other neighbourhoods, Nahalat Binyamin and Geula, into the 'conglomeration' (*Gebilde*) of Tel Aviv (Tel Aviv / Yaffa). However, this is neither mentioned in historical texts about the city nor is it evident in the cityscape. Still, there are two streets bearing those names north of Ahuzat Bayit's former location.

Similarly, neither Florentin, a part of the city characterized by commerce, crafts and industry, nor Mahane Israel (later renamed Kerem HaTeimanim), a Yemenite neighbourhood founded in 1904, nor the settlement of Shapira have as yet become part of the city's official historical narrative (Wokoeck 2). Florentin's association with Tel Aviv as a whole is usually established by its role in the present, that is with reference to its current status as a venue for the young artistic scene and its nightlife appeal. The Yemenite neighbourhood derives its association with the city from its central location near HaCarmel Market. Only Shapira lacks strategies of association and inclusion.[36] However, the extant architecture indicates in all three cases that those places are characterized by divergent histories of formation, usage and population. Furthermore, their extent, boundaries and historical connections to their surroundings are clearly visible.

What processes these neighbourhoods – and others that could not be discussed in this context – will undergo in years to come in terms of furthering their perception as significant contributions to the pluralism inherent in the history of Tel Aviv remains to be seen. However, the fact that the extant architecture and spaces themselves preserve information and keep it accessible, at least in parts, is a prerequisite for saving its history from being forgotten and making re-construction possible.

The examples mentioned above demonstrate that Tel Aviv is more than just the White City. Moreover, they show how their restoration is based on a restrictive process of selection but also that re-construction is fundamentally a dynamic process, always subject to societal discourse, negotiation and change. At the same time, architecture persistently preserves information, even in a state of neglect and in phases when it is not in the forefront of the public's mind.

## BIBLIOGRAPHY

Abel, Günter. 'Das Prinzip Rekonstruktion'. *Das Prinzip Rekonstruktion*. Ed. Uta Hassler and Winfried Nerdinger. Zurich: vdf, Hochschulverlag an der ETH, 2010. 64–75. Print.

Altrock, Uwe. 'Tel Aviv'. *Stadtplanung in Israel und Palästina: Der Friedensprozeß als Neubeginn?* Ed. Uwe Altrock. Berlin: Institut für Stadt- und Regionalplanung ISR, 1998. 3–32. Print.

Altrock, Uwe, Grischa Bertram and Henriette Horni. 'Bürgergesellschaftliches Engagement als Katalysator für Rekonstruktionen'. *Geschichte der Rekonstruktion:Konstruktion der Geschichte*. Ed. Winfried Nerdinger. München: Prestel Verlag, 2010. 148–57. Print.

Assmann, Jan and Tonio Hölscher, eds. *Kollektives Gedächtnis und Kulturelle Identität*. Frankfurt am Main: Suhrkamp, 1988. Print.

Bar Or, Galia et al. *Kibbuz und Bauhaus: Pioniere des Kollektivs*. Leipzig: Spector Books, 2012. Print.

Baus, Ursula. 'Facetten einer Begriffsgeschichte: Rekonstruktion'. *Rekonstruktion in Deutschland: Positionen zu Einem Umstrittenen Thema*. Eds. Michael Braum and Ursula Baus. Basel: Birkhäuser, 2009. 98–105. Print.

Brand, Roy and Ori Scialom, eds. *The Urburb: Patterns of Contemporary Living*. Tel Aviv: Sternthal, 2014. Print.

Confino, Alon and Peter Fritsche. 'Introduction'. *The Work of Memory: New Directions in the Study of German Society and Culture*. Ed. Alon Confino and Peter Fritsche. Illinois: University of Illinois Press, 2002. 1–24, Print.

Cuéllar, Gabriell. 'Ruins under Construction: An Indirect Portrait of Jaffa's Etzel Museum'. Feb. 2013. Web. 2 October 2013. <http://www.forensic-architecture.org/wp-content/uploads/2013/03/Cuellar-Ruins-Under-Construction.pdf>.

Dachs, Gisela. 'Das Deutsche Dorf in Tel Aviv'. *Die Zeit* 8 Jan. 2008. Web. 24 Sep. 2014. <http://www.zeit.de/2007/52/Tel-Aviv-Templer>.

Donna, Batya. 'From Gymnasium to Tower'. *Ariel 77/78* (1989): 93–8. Print.

Efrat, Zvi. 'Bauhausbauten Ohne Bauhaus: Wie die Weiße Moderne zum Volksgut Wurde – und Langsam Ergraute'. *Zeitschrift der Stiftung Bauhaus 2*. Ed. Philip Oswalt. Leipzig: Spector Books, 2011. 6–11. Print.

Feindt, Gregor et al. 'Entangled Memory: Toward a Third Wave in Memory Studies'. *History and Theory* 53 (2014): 24–44. Print.

Fischer, Manfred F. 'Rekonstruktionen – Ein Geschichtlicher Überblick'. in Deutsches Nationalkomitee für Denkmalschutz'. *Rekonstruktion in der Denkmalpflege: Überlegungen, Definitionen, Erfahrungsberichte*. Ed. Juliane von Kirschbaum and Annegret Klein Bonn: Deutsches Nationalkomitee für Denkmalschutz c/o Bundesministerium des Innern, 1998. Print.

Förg, Günther. *Photographs; Bauhaus Tel Aviv – Jerusalem* [on the Occasion of the Exhibition Günther Förg. Photographs. Bauhaus Tel Aviv, Jerusalem, Schillermuseum, Weimar, 15.2.–14.4.2002, Tel Aviv Museum of Art, November – December 2002]. Ostfildern-Ruit: Hatje Cantz, 2002. Print.

Glenk, Helmut. *From Desert Sands to Golden Oranges: The History of the German Templer Settlement of Sarona in Palestine 1871–1947*. Victoria: Trafford, 2005. Print.

Göckede, Regina. 'Das Bauhaus Nach 1933: Migrationen und Semantische Verschiebungen'. *Mythos Bauhaus: Zwischen Selbsterfindung Und Enthistorisierung*. Ed. Anja Baumhoff and Magdalena Droste. Berlin: Reimer, 2009. 276–91. Print.

Hanselmann, Jan Friedrich. 'Vorwort'. *Rekonstruktion in der Denkmalpflege: Texte aus Geschichte und Gegenwart*. Ed. Jan Friedrich Hanselmann. Stuttgart: Fraunhofer IRB Verlag, 2009.

Harpaz, Nathan. *Zionist Architecture and Town Planning: The Building of Tel Aviv (1919–1929)*. West Lafayette, IN: Purdue University Press, 2013. Print.

Heinze-Greenberg, Ita. 'Tel Aviv: Die Erste Jüdische Stadt'. *Europa in Palästina: Die Architekten des Zionistischen Projekt 1902–1923*. Ed. Ita Heinze-Greenberg. Zurich: Gta-Verlag, 2011. 105–29. Print.

Herbert, Gilbert and Silvina Sosnovsky. *Bauhaus on the Carmel and the Crossroads of Empire: Architecture and Planning in Haifa During the British Mandate*. Jerusalem: Yad Izhak Ben-Zvi, 1993. Print.

Herzl, Theodor. *Altneuland*. Leipzig: Seemann, 1902. Print.

Institut für Auslandsbeziehungen, Stuttgart, and Architekturmuseum der Technischen Universität, Munich, eds. *Tel Aviv: Neues Bauen 1930–1939: Fotografien von Irmel Kamp-Bandau*. Tübingen/Berlin: Wasmuth, 1994.

Knufinke, Ulrich. *Bauhaus, Jerusalem: A Guide Book to Modern Architecture (1918–1948)*. Tel Aviv: Bauhaus Center, 2012. Print.

Lerer, Tami. *Sand and Splendor: Eclectic Style Architecture in Tel-Aviv = Nobles Splendeurs Sur Fond De Sable: L'architecture Éclectique À Tel Aviv = Pracht Auf Sanddünen: Eklektische Architektur in Tel Aviv = [ … ]*. Tel Aviv: Bauhaus Center, 2013. Print.

Levin, Michael D. *White City: International Style Architecture in Israel, a Portrait of an Era*. Tel Aviv: Tel Aviv Museum, 1984. Print.

Metzger-Szmuk, Nitza, *Dwelling on the Dunes: Tel Aviv: Modern Movement and Bauhaus Ideals*. Paris: Eclat, 2004. Print.

Minta, Anna, 'Das Herzlia-Gymnasium in Tel Aviv (1909): Synthese Zionistischer Vision und Messianischer Verheißung'. *Israel Bauen: Architektur, Städtebau und Denkmalpolitik Nach der Staatsgründung 1948*. Ed. Anna Minta. Berlin: Reimer, 2004. 365–80. Print.

'Nakbar Map'. *Zochrot*. Web. 5 Nov. 2014 <http://zochrot.org/en/site/nakbaMap>.

Nimrod, Luz, 'The Politics of Sacred Places: Palestinian Identity, Collective Memory, and Resistance in the Hassan Bek Mosque Conflict'. 2008. Web. 2 Oct. 2010. <http://www.indiana.edu/~workshop/colloquia/materials/papers/luz_paper.pdf>.

Rössler, Hans-Christian. 'Der Kampf um die Weiße Stadt'. *Frankfurter Allgemeine Zeitung* 8 May 2013. Web. 20 Sep. 2014 <http://www.faz.net/aktuell/gesellschaft/tel-aviv-der-kampf-um-die-weisse-stadt-12175774.html>.

Rotbard, Sharon. *White City, Black City* (עיר לבנה, עיר שחורה). Tel Aviv: Bavel, 2005. Print.

——. 'Weiße Stadt, Schwarze Stadt'. *Zeitschrift der Stiftung Bauhaus 2*. Ed. Philip Oswalt. Leipzig: Spector Books, 2011. 34–41. Print.

——. *History of South Tel Aviv*. 5 Feb. 2012. Web. 5 Nov. 2014. <http://www.youtube.com/watch?v=3mXUpDeb_Hk>.

——. *White City, Black City: Architecture and War in Tel Aviv and Jaffa*. London: Pluto Press, 2015. Print.

——. *White City, Black City: Architecture and War in Tel Aviv and Jaffa*. Cambridge, MA: MIT Press, 2015. Print.

Rotscheid, Ina. 'Wie das Bauhaus Nach Tel Aviv kam'. *Deutsche Welle*. 30 Mar. 2009. Web. 3 Nov. 2014 <http://www.dw.de/wie-das-bauhaus-nach-tel-aviv-kam/a-4138786>.

'Sarona'. *Israelmagazin*. Web. 24 Sep. 2014. <http://www.israelmagazin.de/israel-orte/tel-aviv-telaviv/sarona>.

'Shalom Meir Tower'. *Wikipedia*. Web. 2 Nov. 2014. <http://en.wikipedia.org/wiki/Shalom_Meir_Tower>.

Stein, Claudia, *Tel Aviv*. Norderstedt: Books on Demand, 2012. Print.

'Tel Aviv / Yaffa'. *Israel Reisen*. Web. 27 Oct. 2014. <http://www.reisen-nach-israel.de/index.php?id=35>.Turner, Judith, and Michael D. Levin. *White City: International Style Architecture in Israel*. Tel Aviv: Tel Aviv Museum, 1984. Print

Veser, Thomas. 'Sanierung der "Weißen Stadt": Instrument Baurechtstransfer: Tel Aviv'. *Bauwelt* 4 (2005). Print.

*White City of Tel Aviv: the Modern Movement.* Web. 3 Nov. 2014. <http://whc.unesco.org/
en/list/1096>.Will, Thomas. 'Projekte des Vergessens? Architektur und Erinnerung
unter den Bedingungen der Moderne'. *Bauten und Orte als Träger von Erinnerung:
Die Erinnerungsdebatte und die Denkmalpflege.* Eds. Hans-Rudolf Meier and Marion
Wohlleben. Zurich: vdf, Hochschlverlag an der ETH, 2000. Print.

Woelke, Miriam. 'Ha'Tachanah: Der alte Bahnhof von Tel Aviv, Teil 1'. *Leben in Jerusalem.*
2 Oct. 2011. Web. 5 Nov. 2014 <http://lebeninjerusalem.blogspot.co.uk/2011/10/
hatachanah-der-alte-bahnhof-von-tel.html>.

Wokoeck, Ursula. 'Hundertjahrfeier in Tel Aviv-Yafo'. *Heinrich Boell Stiftung.* 30 May
2005. Web. 5 Nov. 2014 <http://il.boell.org/sites/default/files/downloads/Tel_Aviv_
centennial_May_2009.pdf>.*Zochrot.* Web. 5 Nov. 2014. <http://zochrot.org/en/place/
al-manshiyya-neighborhood-yaffa>.

## NOTES

1   See the *Duden* definition of *rekonstruieren*. Retrieved from: <http://www.duden.
    de/rechtschreibung/rekonstruieren> 4 Nov.2013. Synonyms given here include
    *nachbilden* (recreate), *nachformen* (shape after), *nachgestalten* (design after), *wieder
    errichten* (re-erect), *aufleben lassen* (revive), *wiedergeben* (reproduce).

2   For an introduction into the concept of reconstruction see for example Fischer 7–15
    and Hanselmann 1. Architecture historian Winfried Nerdinger categorically casts
    reconstructions in a favourable light and argues, among other things, that they have
    always been common practice, see Nerdinger 10–14. The exhibition at TU München's
    architecture museum and the catalogue are setting standards for topics of research in
    this field. They have received considerable attention and critiques from reviewers.

3   See, among others, Hanselmann 5.

4   'In den allermeisten Fällen nur nach einem Erscheinungsbild gerufen [wird] – und
    nicht nach einem historischen Raumgefüge, einer historischen Funktion oder allem
    zusammen in einer Spurensuche über die Zeit hinweg'. Günter Abel approaches
    the concept from the viewpoint of philosophy and considers reconstructions in
    architecture 'as constructs of translation and interpretation' (*als Übersetzungs- und
    Interpretationskonstrukte*), among other things. Abel 68.

5   On the motivation of proponents of reconstruction since 1975 see Altrock, Bertram
    and Horni.

6   Basic reading: Assmann and Hölscher. For the newest trends in research, see Feindt.
    For the research project, Confino and Fritzsche's action-orientated definition is
    relevant. They view memory 'as a symbolic representation of the past embedded in
    social action, [ … ] as a set of practices and interventions' (Confino and Fritsche 5).

7   This is meant to be considered in relation to the publicness of built space. Generally
    speaking, there are so many media capable of conserving the memory of a building
    and thus a time period or event that it cannot be reasonably said that it is possible to
    'forget' the information in terms of it being lost or becoming inaccessible.

8   As a basis for the following summary see Heinze-Greenberg, also detailing the
    requisite purchase of land, financing and the situation in Jaffa.

9   The association had been founded as early as 1907, see Heinze-Greenberg 107.

10  'altneues Hebräertum kommunizieren' (Heinze-Greenberg 117).

11    For this phase, see Harpaz.

12    There are several publications on the architecture of this period. See for example
the extensive catalogue Metzger-Szmuk, in which numerous biographical notes on
architects can be found and a systematic overview of building types is undertaken.

13    Uwe Altrock remarked in 1998 that these had been documented in a municipal
conservation inventory but that there was no 'strong evidence against continued
improper handling' (*starke Handhabe gegen einen weiteren unsachgemäßen Umgang*)
(Altrock 26, footnote 25). The current state of buildings known to the author of this
paper suggests that this assessment is still valid. Also, there has been no scientific
publication on the individual buildings, their history, the settlement's development or
what is done with them today.

14    See for example the article 'Shalom Meir Tower', as well as Donna. According to Anna
Minta, the Herzliya Gymnasium was 'the first official manifestation of Zionism with
the intention of creating a new-old geographical and spiritual home in Palestine
using methods of settlement and cultural work' (*eine erste offizielle Manifestation des
Zionismus, mit den Methoden der Siedlungs- und Kulturarbeit in Palästina eine neu-alte
geographische wie auch spirituelle Heimat aufzubauen*). See Minta 380.

15    On the model, no indication is to be found of what stage in the city's development
process it is supposed to depict. The fact that it is associated with the exhibition on
'Tel Aviv's Modern Movement', and the UNESCO World Heritage Site and the presence
of some high-rises to the south of the city, hotels near the beach and the Dizengoff
Center suggest that it is a view of the city as of the early 2000s.

16    In addition, the Bauhaus Center has published documentation on a selection of
buildings in their current condition with some basic notes on their construction (Lerer).

17    To mention only two examples: there is a study by Israeli architect Zvi Efrat about
architecture in Israel between 1948 and 1973, which is not yet available in German
and only partly in English. However, Basel's architecture museum hosted an exhibition
curated by him from 23 October 2011 to 9 April 2012 entitled *The Object of Zionism:
Architektur und Stadt 1948–1973*. No catalogue was published. A selection of texts
and pictures from the project can be found on the internet: <http://www.efrat-
kowalsky.co.il/texts/the-israeli-project/>. See also a brief overview on 1950s and 60s
architecture in Efrat. During the Venice Biennale of Architecture in 2014, there was
a focus on the so-called 'new towns' established since the founding of the state. By
examining their development up until the present, it was possible to close a gap in
this field of research. See the catalogue by Brand and Scialom.

18    Apart from them, there are many more examples to be found in other cities as well as
in kibbutzim. For the case of Jerusalem, see Knufinke, for Haifa: Herbert and Sosnovsky.
As yet, there is no comprehensive documentation of those buildings available,
however. For kibbutzim: Bar Or et al.

19    See the two catalogues Levin; and Turner and Levin.

20    The exhibition originally opened at the Institut für Auslandsbeziehungen's Forum für
Kulturaustausch ('cultural exchange forum') between 9 June and 25 July 1993,
then at Bauhaus-Archiv, Museum für Gestaltung (Berlin) from 28 September to
28 November 1993 and at Ifa-Galerie (Bonn) between 4 May and 11 June 1994.
After that, it went on tour to Tel Aviv (1994), Vienna and Brno (1995), Brussels and
Leuven (1996), Coimbra, Frankfurt and Zurich (1997), Grenoble, Saint-Etienne, Austin
and New York (1998), Tallinn (1999), Prague and Budapest (2000) and Ankara (2001).
Special thanks for this information goes to Stefanie Alber of ifa-Galerie Stuttgart/
Institut für Auslandsbeziehungen (ifa), email of 8 October 2014.

21   As of July 2013.

22   The first exhibition was shown from 20 May to 7 August 2004 at the Helena Rubenstein Pavilion for Contemporary Art, later, among others, in Canada, Austria, Germany, Switzerland and finally under the title *White City: Bauhaus Architecture in Tel Aviv* at the State Hermitage Museum Saint Petersburg (11 June until 13 September 2013) and as *The White City – Tel Aviv's Modern Movement* at the Museum of Finnish Architecture in Helsinki (12 February to 30 April 2014).

23   Apart from the exhibitions mentioned, this becomes particularly apparent in photographic collections (one example is Förg) and on guided city tours.

24   'Bauhaus' is probably the term most often used in relation to this architecture, be it in everyday conversation, media reports or scientific discourse. To mention just one example, see Rotscheid. The problems arising from this attribution cannot be fully discussed here. Suffice it to say that it does not stand up to the state of research into the actual buildings in terms of history of architecture and also lumps together quite a number of different cues from European modernism and thus diverging experiences and biographical backgrounds of the architects involved. For insights into the application of the Bauhaus label to Tel Aviv and Israel, see Göckede, of special note: 'Die Israelische DissemiNation des Bauhaus', 285–90.

25   It would be desirable both in terms of research and of the architecture's perception in general to examine the city comparatively in the context of the international modernist movement.

26   Sharon Golan of Tel Aviv's municipal monument conservation office made reference to this aspect of the city's self-conception in a call on 4 October 2013. Another factor is the necessity to provide shelters and make the buildings earthquake-proof. The increasingly rampant gentrification resulting from the renovation is not being considered in this context and unfortunately cannot be discussed in depth in this paper.

27   Rotbard 2005. Also now published in the UK by Pluto Press as *White City, Black City: Architecture and War in Tel Aviv and Jaffa* and in the US by MIT Press as *White City, Black City: Architecture and War in Tel Aviv and Jaffa*. Thanks go to the author, who forwarded this information in an email of 7 September 2014. More insights, unfortunately in the form of polemic retribution lacking an empirical basis and inadequate to the significance of his research, are provided by Rotbard 2011.

28   Al-Jammasin al-Gharbi (located in the north, Pinkas St. west of Namir Road), al-Mas'udiyya (Summayl) (located in the west of the city along Gabirol Street, south of Jabotinsky Street), al-Shaykh Muwannis (located in the north of the city within today's Tel Aviv University grounds) and al-Manshiyya (located to the north of Jaffa, along the Mediterranean). Of all these villages, the al-Manshiyya neighbourhood has seen the most extensive documentation (*Zochrot*).

29   Other than that, Zochrot lists three more Arab settlements – and Jaffa – for the Tel Aviv area. As they are located outside the urban area in question, however, they are not given here. See 'Nakbar Map'.

30   For further information on its significance for the collective memory see Nimrod.

31   This is not to say that the place's history is forgotten, seeing how not only the texts selected here make reference to it, but also NGOs like Zochrot.

32   Here, reference might also be made to other former Arab settlements as well as a Muslim cemetery now built over on the grounds of today's Independence Park next to the Hilton Hotel as well as their presence (or lack thereof) in the public space, but this would go beyond the scope of this text.

33  Uwe Altrock states that, in the case of these settlements, 'the Christian and Jewish settlers' European provenance was clearly recognizable in terms of architecture (and partially remains so to this day)' (*der europäische Ursprung der christlichen und jüdischen Siedler baulich deutlich ablesbar war (und bis heute teilweise geblieben ist)*) (Altrock 3).

34  Another example is the extensive renovation work performed on 37 buildings of the former Templer settlement of Sarona considered to be the 'greatest conservation project ever conducted in Tel Aviv' (*größte, jemals durchgeführte Denkmalpflege-Projekt in Tel Aviv*) ('Sarona') and during the course of which, among other things, five buildings were completely relocated. Concerning the place's history, see Glenk or Dachs. The neighbourhood re-opened in 2014. The buildings house shops, galleries and restaurants, as it is primarily intended as a recreational destination. Here too, panels inform about its history. Considering the high-rises surrounding it, a sharp historical contrast between the early German Templer settlers and contemporary, modern Israel is evident.

35  Before and concurrently with Ahuzat Bayit, other settlements were founded outside Jaffa, for instance Neve Shalom, Mahane Yehuda (later renamed Kerem HaTeimanim), Mahane Yosef, Shaarei Ahva and Ohel Moshe (Wokoeck 6).

36  The area tends to be held in low esteem by the public for its location near the New Central Bus Station and the great number of legal and illegal migrants living there. For its history, see Rotbard 2012.

# Index